LECTURES ON
GAME THEORY

UNDERGROUND CLASSICS IN ECONOMICS

Consulting Editors
Kenneth Arrow, James Heckman, Joseph Pechman,
Thomas Sargent, and Robert Solow

Progress in economics takes place in many different forums. Many of these forums are not "above ground," in the sense of published work freely available. They are in the form of unpublished notes, dissertations, government reports, and lectures. This series is dedicated to making the best of this underground literature more widely available to libraries, scholars, and their students. In so doing it will serve to fill a large gap in the contemporary economic literature.

Entries in this series can be on any topic of interest to economists, and they may occasionally be in rough or unfinished form. The only criteria are that the works have been influential, widely cited, of exceptional excellence, and, for whatever reason, never before published.

Other Titles in This Series

Notes on the Theory of Choice, David M. Kreps

Functional Form and Utility: A Review of Consumer Demand Theory, Arthur S. Goldberger

Behind the Diffusion Curve: Theoretical and Applied Contributions to the Microeconomics of Technology Adoption, Paul A. David

The Allocation of Scarce Resources: Experimental Economics and the Problem of Allocating Airport Slots, David M. Grether, R. Mark Isaac, and Charles R. Plott

LECTURES ON
GAME THEORY

ROBERT J. AUMANN

Routledge
Taylor & Francis Group

LONDON AND NEW YORK

First published 1989 by Westview Press, Inc.

Published 2018 by Routledge
52 Vanderbilt Avenue, New York, NY 10017
2 Park Square, Milton Park, Abingdon, Oxon OX14 4RN

Routledge is an imprint of the Taylor & Francis Group, an informa business

Copyright © 1989 Taylor & Francis

Library of Congress Cataloging-in-Publication Data
Aumann, Robert J.
   Lectures on game theory / by Robert J. Aumann
   p.   cm. -- (Underground classics in economics)
   ISBN 0-8133-7578-9
   1. Game theory.   2. Economics, Mathematical.   I. Title.
II. Series.
HB144.A95 1989
330'.01'5193--dc19   88-11097
                                                     CIP

ISBN 13: 978-0-367-01217-5 (hbk)
ISBN 13: 978-0-367-16204-7 (pbk)

Contents

Preface

This book is a collection of lectures given at the Economics Department at Stanford University during the fall and winter quarters of 1975-1976. I am very grateful to José Córdoba and Haruo Imai, who took the notes and wrote up the lectures, and to Martin Osborne, who provided the solutions to the exercises.

Game theory has progressed tremendously in the thirteen years since these lectures were given. Yet this material, which represents a first course in the subject, has not been superceded. It is quite possible to give a course based on these lectures today, although undoubtedly, here and there some later developments could be worked in. Even at the time, the material represented a selection; it was not feasible to cover all the important areas in one two-quarter course.

The level of mathematical sophistication of Chapters 1-7 is not high; it should be easily accessible to most upper-level undergraduates and graduate students majoring in economics, operations research, statistics, mathematics, and other quantitatively or formally inclined majors. Chapter 8 is a little more sophisticated.

Before publication, the material was briefly reviewed, and a few minor corrections were made. It is, of course, quite likely that errors remain. Readers are invited to inform me of any that they find.

To avoid introducing new errors, and to speed the publication process, the material was not retyped; instead, corrections were made on the original typescript. This necessitated leaving some gaps in the lines. Moreover, as the observant reader will notice, the typewriter used for the corrections was not the same as the original. This is, perhaps, not altogether bad---the different font signals the new material, much like reconstructions are consciously differentiated at archaeological sites. In any case, I beg the readers' indulgence.

I would like to thank Stanford University, particularly the economics section of its Institute for Mathematical Studies in the Social Sciences and its director, Professor Mordecai Kurz, for creating the opportunity to give these lectures and for arranging to have the notes taken.

Robert J. Aumann
Jerusalem, Israel
Tishri 5749 (September 1988)

Chapter 1: Zermelo's Theorem

Game theory is a theory of rational behavior of people with non-identical interests. Its area of application extends considerably beyond games in the usual sense--it includes, for example, economics, politics, and war. By the term "game" we mean any such situation, defined by some set of "rules." The term "play" refers to a particular occurrence of a game. Thus chess is a game, and several plays of chess took place in the summer of 1972 between Fisher and Spassky.

We begin with Zermelo's theorem on chess.

1.1   Theorem (Zermelo [1912]): In chess either white can force a win, or black can force a win, or both sides can force at least a draw.

Proof: We will prove the result for a family of games that includes chess. Each game in this family is characterized by: (1) a position in chess, (2) an indication of "who must move" (black or white), and (3) a positive integer n (with the understanding that if the game does not end in mate or draw within n single moves at most, then it is declared a draw). (Chess is a member of this family because the number

---

Based on lectures delivered at Stanford University in the fall of 1975 and winter of 1976. Notes taken by Haruo Imai, José Córdoba, and Martin J. Osborne.

of moves in chess is bounded (by the rule whereby a play of chess ends when the same position is repeated three times).)

We prove the result by induction on n. The reason for using the larger family is that it strengthens the inductive hypothesis and so makes the inductive proof possible. This is typical of inductive proofs.

Suppose n = 1. If black moves, black can either mate on that move, or he cannot; in the first case black can force a win, and in the second case, both players force a draw. Similarly for white. Now assume the theorem is correct for all n ≤ m - 1. We wish to deduce the theorem for n = m. Without loss of generality (henceforth abbreviated w.l.o.g.) suppose black moves first. By the induction hypothesis, after black has made the first move, either black can force a win or white can force a win or both can force at least a draw. In other words, with each move by black, designated by p, there is associated a letter f(p) that may be b, w, or d (b, w and d stand respectively for "black can force a win," "white can force a win," and "both can force at least a draw"). Then there arise three mutually exclusive and exhaustive cases:

(1) If there is a move p of black such that f(p) = b, then black can force a win in the original game.

(2) If for all p, f(p) = w, then white can force a win in the original game.

(3) Otherwise, there is no p for which $f(p) = b$, but there is
a p for which $f(p) = d$. Hence black can force at least a draw, and
so can white.

This completes the proof of the theorem.

We now introduce a concept of fundamental importance in game
theory, that of strategy. By the term strategy, we mean a complete plan
for playing a game (for one player), taking all contingencies into
account, including what all other players might do in the course of the
play.

For example, in Tic-Tac-Toe the first player has at most 5 moves,
and for each move there are at most 9 possibilities. Nevertheless he
has far more than 45 strategies. For a strategy is a complete plan,
and the number of possibilities in a complete plan which covers only
the first two moves of the first player is already 504. This is because
for the first move there are 9 possibilities, and for each of the 8
possible responses of the second player, player 1 has 7 choices for his
second move.

In terms of strategies, Zermelo's theorem is illustrated in the
three tables on page 4. The rows represent strategies of white, and the
columns strategies of black. The numbers 1,2,... index the strategies.
To each pair of strategies of white and black, there corresponds one of
the letters w, b and d. If white can force a win then there exists a
strategy of white which, no matter which strategy black plays, assures
w. If black can force a win, then there exists a strategy of black

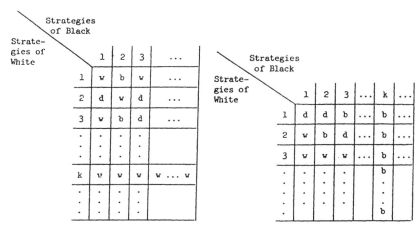

White can force a win if and only if there is a row k that is filled with w's.

Black can force a win if and only there is a column k that is filled with b's.

	1	2	3	...	k	...
1	w	d	b	...	d	...
2	d	b	b	...	b	...
3	w	w	b	...	b	...
.	.	.	.		b or d	
.	.	.	.			
.	.	.	.			
k'	w	d	w or d		d	w or d
.	.	.	.		b or d	
.	.	.	.			
.	.	.	.			

Both can force at least a draw if and only if there are a column k that does not have a w and a row k' that does not have a b.

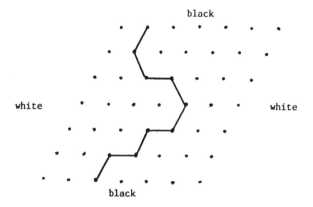

A "connected" set linking black's sides in Hex.

which, no matter which strategy white plays, assures b. If both can

force at least a draw, then there is no strategy which assures w

to white and no strategy which assures b to black, but there are

strategies of white and black which assure at least d for both white

and black.

1.2   **Example** (Hex): The following rules define the game of "Hex."

There are two players, white and black. The board consists of a rhombus

of dots, with angles of 60 and 120 degrees, as in the drawing above; two

opposite sides are labelled "white," the other two "black." (For a

challenging game, the board should be larger, e.g., 11x11). Call two

dots <u>adjacent</u> if each is closest to the other; thus each dot in the

interior has six adjacent dots. The idea is for each player to try to

link up his two sides. White moves first, by "capturing" a dot anywhere

in the rhombus. The players then take turns capturing previously

uncaptured dots (NOT necessarily adjacent to previously captured dots).

A player wins if the dots he has captured include a "connected" set of

adjacent dots linking his two sides (as in the drawing).

Clearly, Zermelo's theorem applies to this game. Also,

a little thought will show that a draw is impossible in this game. Thus either white can force a win or black can force a win. We now prove that in fact white can force a win. To this end, we show that if black could force a win, then white could force a win.

Let us define <u>Reversed Hex</u> to be the same as Hex, except that black moves first. If black could force a win in Hex, then white could force a win in Reversed Hex. Consider now a play of Hex. White can play by making an arbitrary move to begin with, subsequently ignoring that move and playing to win as if he were in Reversed Hex. If at any time his strategy dictates occupying the point he occupied on his first move, then he can simply occupy another arbitrary point. This will lead to a win for him, which is a contradiction. Thus it is impossible for black to force a win in Hex; hence white can force a win.

Note that this is simply a proof of the <u>existence</u> of a winning strategy for white; even for boards of moderate size (say 12 × 12 or 13 × 13) no winning strategy for white is actually known.

1.3   <u>Example</u> (Kriegsspiel): Consider the game known as "Kriegsspiel," in which black   and white play chess separately without knowing each other's position; each is informed when a move he proposes is illegal because of the positions of the pieces of the other player. For this game, the proof of Zermelo's theorem given above cannot be applied, and in fact the theorem itself probably does not hold. The difference between chess and Kriegsspiel lies in the fact that at each stage in chess every move made up to that stage is known by both players, while this is not true in Kriegsspiel. This makes the inductive step in the proof of Zermelo's theorem invalid, because the player whose turn it is to move does not know to what position his move will lead. Technically, chess is a game of perfect information, whereas Kriegsspiel is not.

Chapter 2: Noncooperative Games

In the games we have discussed up to now there are 2 players whose interests are completely opposed. It is clear that when there are more than 2 players, they cannot have completely opposed interests. This motivates the following definition.

2.1   Definition: A game is strictly competitive if it has two players (1 and 2) and for any two possible outcomes x and y, if 1 prefers x to y, then 2 prefers y to x.

In a strictly competitive game, we can assign numbers to the outcomes such that a higher number corresponds to an outcome that player 1 prefers. If we do this in chess, then Zermelo's theorem asserts that there is a number v such that white can guarantee that his payoff will be at least v, and black can guarantee that white's payoff will be no more than v. This motivates the following definition.

2.2   Definition: A number v is said to be the minimax value of a strictly competitive game if player 1 can guarantee that his payoff will be at least v, and player 2 can guarantee that the payoff of player 1 will be no more than v.

Not all strictly competitive games have minimax values. For example, the game "matching pennies," in which the payoffs to 1 are those given in the following table has no minimax value.

		Player 2	
		Strategy 1	Strategy 2
Player 1	Strategy 1	1	-1
	Strategy 2	-1	1

We would now like to generalize our considerations to games that are not necessarily strictly competitive and may have more than 2 players.

2.3   Definition: A game G (in strategic form) consists of:

   (1)   a set N (the players);

   (2)   for each player i, a set S^i (the strategies of i); and

   (3)   for each player i, a function $h^i: \underset{i \in N}{\times} S^i \to \mathbb{R}$ (the payoff function of i).

2.4   Remark: The term "strategic form" is used to indicate that we have abstracted from individual moves and are looking only at strategies.

   If $s \in \underset{i \in N}{\times} S^i$ (i.e. s is an n-tuple of strategies) and $t^i \in S^i$ (i.e. t^i is a strategy of i), write $s|t^i$ for the n-tuple of strategies which is the same as s except that t^i is substituted for i's strategy s^i in s.

2.5   Definition: An equilibrium point of G is an n-tuple s of strategies such that for any player i and for any strategy t^i of i, $h^i(s|t^i) \leq h^i(s)$.

A two-person zero-sum game is a game G with n = 2 such that
for all strategy pairs s, we have $h^1(s) + h^2(s) = 0$. Clearly, a two-
person 0-sum game is strictly competitive. In such a game, a pair s
is an equilibrium point if for any t^1 and t^2, $h^1(s|t^1) \leq h^1(s)$ and
$h^2(s|t^2) \leq h^2(s) = -h^1(s)$; or, if for any t^1 and t^2,

$$h^1(t^1,s^2) \leq h^1(s^1,s^2) \leq h^1(s^1,t^2) .$$

This means that $h^1(s^1,s^2)$ is the minimax value of the game. Thus we
see that a two-person 0-sum game has a minimax value if and only if it
has an equilibrium point. We now wish to prove a proposition that
connects the existence of an equilibrium point (or equivalently, of a
minimax value) in a two-person 0-sum game to what is called the "minimax
property."

2.6   Definition: A subset of a Euclidean space is said to be compact
if it is bounded and closed.

2.7   Remark: A real-valued continuous function on a compact set
attains its maximum and its minimum (the proof is left to the reader).

2.8   Proposition: Let G be a two-person zero-sum game. Assume that
the s^i are compact subsets of Euclidean spaces and the h^i are continuous. Then a necessary and sufficient condition for the existence of
an equilibrium point in G is that

$$\max_{s^1} \min_{s^2} h^1(s^1,s^2) = \min_{s^2} \max_{s^1} h^1(s^1,s^2) .$$

2.9   Remark: The quantity max min $h^1(s^1,s^2)$ represents the largest amount that player 1 can guarantee to himself by playing an appropriate strategy. Similarly, min max $h^1(s^1,s^2)$ represents the smallest amount such that player 2 can guarantee that player 1 will not obtain more than that amount. In "matching pennies," which has no minimax value, these amounts are different: we have max min = -1 and min max = 1.

2.10   Remark: The compactness and continuity of the payoff functions h^i are needed to assure that max min and min max exist.

2.11   Remark: The Cartesian product of compact sets is compact (the proof is left to the reader).

2.12   Remark: Let h be continuous on $S^1 \times S^2$ and let S^1 and S^2 be compact. Then $\min_{s^2} h(s^1,s^2)$ is a continuous function of s^1. (The proof is left to the reader.)

   Proof of Proposition: We first assert that always

(1) $$\max_{s^1} \min_{s^2} h^1 \leq \min_{s^2} \max_{s^1} h^1 \quad .$$

Indeed, for any s^1 and s^2 we have

         $$\max_{s^1} h^1(s^1,s^2) \geq h^1(s^1,s^2) \quad .$$

Taking the minimum over s^2 on both sides, we deduce

$$\min_{s^2} \max_{s^1} h^1(s^1,s^2) \geq \min_{s^2} h^1(s^1,s^2) \quad .$$

Since this holds for all s^1, it holds also for an s^1 at which the right hand side attains its maximum; hence

$$\min_{s^2} \max_{s^1} h^1(s^1,s^2) \geq \max_{s^1} \min_{s^2} h^1(s^1,s^2) \quad .$$

This completes the proof of (1).

Assume now that s_0 is an equilibrium point; i.e. for all s^1 and s^2,

$$h^1(s^1,s_0^2) \leq h^1(s_0^1,s_0^2) = v \leq h^1(s_0^1,s^2) \quad .$$

Then for all s^1 and s^2,

$$\max_{s^1} \min_{s^2} h^1 \geq \min_{s^2} h^1(s_0^1,s^2) \geq h^1(s_0^1,s_0^2) = v \geq \max_{s^1} h^1(s^1,s_0^2)$$

$$\geq \min_{s^2} \max_{s^1} h^1(s^1,s^2) \quad .$$

Together with (1), this yields

$$\max_{s^1} \min_{s^2} h^1 = \min_{s^2} \max_{s^1} h^1 \quad .$$

Finally assume, conversely, that

$$\max_{s^1} \min_{s^2} h^1 = \min_{s^2} \max_{s^1} h^1 \quad .$$

Suppose the maximum on the left is achieved at s_0^1; i.e.

$$\max_{s^1} \min_{s^2} h^1(s^1,s^2) = \min_{s^2} h^1(s_0^1,s^2) \quad .$$

Similarly, let

$$\min_{s^2} \max_{s^1} h^1(s^1,s^2) = \max_{s^1} h^1(s^1,s_0^2) \quad .$$

For all s^1 and s^2, we therefore have

$$h^1(s_0^1,s^2) \geq \min_{s^2} h^1(s_0^1,s^2) = \max_{s^1} h^1(s^1,s_0^2) \geq h^1(s^1,s_0^2) \quad .$$

Substituting $s = s_0$ we get

$$h^1(s_0^1,s_0^2) = \max_{s^1} \min_{s^2} h^1 = \min_{s^2} \max_{s^1} h^1 \quad ,$$

and hence

$$h^1(s^1,s_0^2) \leq h^1(s_0^1,s_0^2) \leq h^1(s_0^1,s^2) \quad .$$

Thus s_0 is an equilibrium point, and the proof of the proposition is complete.

We introduce now the concept of the "mixed extension of a game."

2.13 **Example**: Consider the game "matching pennies," introduced above. The interpretation is that each player shows the other one side of a coin. If both players show the same side then player 1 wins. If not, player 2 wins. We have seen that this game has no minimax value, or, equivalently, no equilibrium point. As a consequence, no playing system can be sustained by a player, since the other player can outguess it and win. So each player ends up playing at random; i.e. each player goes to a corner, tosses the coin and shows the side thus determined. Playing at random in this way is equivalent to choosing strategies 1 and 2 each with probability 1/2. If a player plays at random, his expected payoff is 0, no matter what the strategy chosen by the other player is. We notice that random play expands the possibilities for strategy choices. Each player can now choose among a continuum of strategies--a continuum that we may represent by the unit interval [0,1]. We have a new game defined by:

$$N = \{1,2\} \quad ;$$

$$S^1 = \{p: \ 0 \leq p \leq 1\} \quad , \quad S^2 = \{q: \ 0 \leq q \leq 1\} \quad ;$$

and $H^1(p,q) = pq \cdot 1 - (1 - p)q \cdot 1 + (1 - p)(1 - q)1 - (1 - q)p \cdot 1$

$$= (1 - 2p)(1 - 2q) \quad .$$

This new game is called the <u>mixed extension</u> of "matching pennies."

2.14   Definition: A _mixed strategy_ in a game   G   is a strategy in the mixed extension of the game.

Sometimes, when we want to emphasize that we are dealing with strategies in   G, rather than mixed strategies, we call them _pure strategies_.

2.15   Assertion: The mixed extension of "matching pennies" has a unique equilibrium point.   The equilibrium point is $p = 1/2$, $q = 1/2$.

Proof: The existence of an equilibrium point (e.p.) follows directly from the fact that

$$H^1(p, \tfrac{1}{2}) = H^1(\tfrac{1}{2}, \tfrac{1}{2}) = H^1(\tfrac{1}{2}, q) \quad .$$

Since $\min_q H^1(p,q)$ is a function of   p   with a unique maximum, the e.p. is unique.

The following is another example of a zero-sum game with no pure strategy equilibria, but with an equilibrium point in the mixed extension.

2.16   Example: Consider a game with payoff matrix as below.

		s^2_1 (q)	s^2_2 (1 - q)
s^1_1	(p)	1	3
s^1_2	(1 - p)	4	2

We see that there is no pure strategy equilibrium in this game. Let p and q be the mixed strategies of players 1 and 2. Then

$$H^1(p,q) = pq \cdot 1 + p(1-q)3 + (1-p)q \cdot 4 + (1-p)(1-q)2$$

$$= -4pq + p + 2q + 2$$

$$= -(1-2p)(\frac{1}{2} - 2q) + \frac{5}{2} \ .$$

From this formula it can be seen that $p = 1/2$, $q = 1/4$ is an equilibrium point. We can also compute it by a direct method: namely by computing p such that $\min_q H^1(p,q)$ is maximum (and q such that $\max_p H^1(p,q)$ is minimum) and verifying that $\max_p \min_q H^1(p,q) = \min_q \max_p H^1(p,q)$. We have

$$\min_q H^1(p,q) = \begin{cases} p + 2 & \text{if } p \le \frac{1}{2} \\ \\ -3p + 4 & \text{if } p \ge \frac{1}{2} \end{cases} \ .$$

The graph of $\min_q H^1(p,q)$ looks as follows:

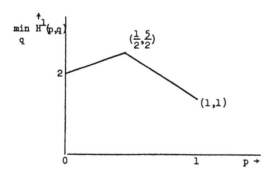

From this it is apparent that $\max_{p} \min_{q} H^1(p,q) = 5/2$. Similarly,

$$\max_{p} H^1(p,q) = \begin{cases} -2q + 3 & \text{if } q \leq \frac{1}{4} \\ 2q + 2 & \text{if } q \geq \frac{1}{4} \end{cases} \quad ;$$

the graph of $\max_{p} H^1(p,q)$ thus looks as follows:

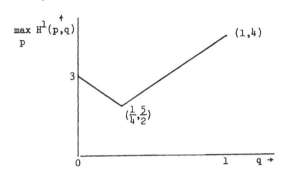

Hence $\min_{q} \max_{p} H^1(p,q) = 5/2$. Since $\max_{p} \min_{q}$ is achieved when $p = 1/2$, and $\min_{q} \max_{p} H^1(p,q)$ is achieved when $q = 1/4$, it follows that $(1/2, 1/4)$ is an equilibrium point.

In the games we are going to discuss next, the interests of the players are not completely divergent. They are called non-zero-sum games.

2.17 Example: Consider the game with payoff matrix as below. This game has two pure strategy equilibrium points (s_1^1, s_1^2) and (s_2^1, s_2^2). These points remain equilibrium points in the mixed extension of the game; they correspond to $p = 1$, $q = 1$ and $p = 0$, $q = 0$. There is however

		s^2_1	s^2_2
		(q)	(1 - q)
s^1_1	(p)	1,1	0,0
s^1_2	(1 - p)	0,0	1,1

an additional equilibrium point, $p = 1/2$, $q = 1/2$, since

$$H^1(p,\tfrac{1}{2}) = \tfrac{1}{2} \qquad \text{for all } p$$

and

$$H^2(\tfrac{1}{2},q) = \tfrac{1}{2} \qquad \text{for all } q \ .$$

The expected payoff associated with this equilibrium point is $(1/2,1/2)$ which is smaller than the payoff corresponding to the two other equilibria. This result provides a case for two possible interpretations of an equilibrium point. First, an e.p. may be interpreted as a self-enforcing agreement. Once such an agreement is written down, it is of no advantage to either one of the two sides to violate it. For instance, if the two players agree to choose the point (s^1_1,s^2_1), neither has any advantage in moving away from it. An agreement that is not an equilibrium point will be violated because there is an incentive to do so and there is no enforcement mechanism. This interpretation is relevant in situations like international treaties and illegal collusions on constrained trade. Alternatively, an e.p. can be interpreted as a natural outcome when there is no possibility of communication between the players, but somehow the e.p. in question is "prominent" or "natural;" i.e. each player has reason to believe that the other one will play in accordance with it. In the case above the e.p. $(1/2, 1/2)$ is such an outcome since the players cannot agree on either (s_1^1, s_1^2) or (s_2^1, s_2^2).

2.18   Example: Consider the game with payoff matrix as below.

	s_1^2 (q)	s_2^2 (1 - q)
s_1^1 (p)	1,1	0,0
s_2^1 (1 - p)	0,0	2,2

The mixed extension of this game has three equilibria:

$$p = 1 \ , \quad q = 1 \ \Big\} \ \text{pure strategies}$$
$$p = 0 \ , \quad q = 0$$

$$p = \frac{2}{3} \ , \quad q = \frac{2}{3} \quad \text{mixed strategies} \ .$$

Note that the mixed strategy equilibrium yields a payoff that is worse for both players than either of the pure strategy e.p.'s. So it seems unlikely that this e.p. will be chosen even when communication is impossible. However, in this case it seems just as unlikely that the pure e.p. (1,1) will be chosen, since the e.p. (2,2) is better for both players than all other e.p.'s.

2.19   Example: In the two above examples, we dealt with purely cooperative situations. Consider now a bargaining situation.

	s_1^2 (q)	s_2^2 (1 − q)
s_1^1 (p)	2,1	0,0
s_2^1 (1 − p)	0,0	1,2

This game is sometimes called the "Battle of the Sexes." (We may imagine
that the husband prefers vacationing in the mountains while the wife
prefers vacationing by the seaside, but both prefer a vacation to staying
at home.) There exist two pure strategy equilibria $p = 1$, $q = 1$ and
$p = 0$, $q = 0$, and one mixed strategy equilibrium $p = 1/3$, $q = 2/3$. The
mixed strategies outcome $(1/3, 2/3)$ makes the two players equally well
(or badly) off, but is not efficient. It is dominated by both of the
two other equilibria: both players are better off when moving to either
pure strategy equilibrium. Here, the main problem is one of bargaining.
Both players have an incentive to reach an agreement through bargaining,
since it would ensure one of the two pure strategy equilibria. However,
if an agreement cannot be reached (either because the bargaining process
is unsuccessful or because communication is impossible), then the mixed
strategy equilibrium is the natural outcome.

2.20   Example (Prisoners' Dilemma): Two prisoners are arrested by the
police, but there is not enough evidence to convict them. The police ask
them to give evidence against each other.

There are three possible cases (two out of the four situations are symmetric). If one cooperates with the police and the other does not, the outcome is best for the one who cooperates (he gets freedom plus side advantages, such as a job and new identity) and worst for the other (he stays in jail under hard conditions). If both turn state's evidence, they will not be freed but will benefit from good treatment in jail. If both do not, both will be freed but cannot get side advantages.

The situation is expressed by the following payoff matrix:

Prisoner 1 \ Prisoner 2	not cooperating with the police	cooperating with the police
not cooperating with the police	4,4	0,5
cooperating with the police	5,0	1,1

(1,1) is seen to be the only equilibrium in either pure or mixed strategies. Some questions arise here. Why is (4,4) not called an equilibrium in game theory, since (4,4) dominates (1,1)? According to the logic of game theory, though (4,4) may be a "good" outcome, it is not self-enforcing and so not "stable."

Note that in this example, one does not need the notion of equilibrium to justify the conclusion that each prisoner will turn state's evidence. That is because turning state's evidence is best for each prisoner no matter what the other one does. Thus unlike in the previous examples, no assumptions at all about the other player's behavior are needed to justify the conclusion.

2.21   <u>Example</u>: Consider the game with payoff matrix below.

		s_1^2	s_2^2
		(β)	$(1 - \beta)$
s_1^1	(α)	1,0	0,1
s_2^1	$(1 - \alpha)$	$\frac{1}{2},\frac{1}{3}$	1,0

Like matching pennies, this game has no pure strategy equilibrium; unlike matching pennies it is not strictly competitive. Now, let mixed strategies α, β be as in the table.

   <u>Assertion</u>: If (α,β) is an equilibrium point, then $0 < \alpha < 1$, $0 < \beta < 1$.

   <u>Proof</u>: Suppose not; for example, let $\alpha = 0$. Then $\beta = 1$; but this is not an equilibrium point. Similarly for the other case.

   If player 2 plays s_1^2 he will get $(1/3)(1 - \alpha)$, and if he plays s_2^2 he will get α. So for an equilibrium $(1/3)(1 - \alpha) = \alpha$, given the claim of the assertion. Thus $\alpha = 1/4$. A similar calculation for 1 yields $\beta = 2/3$. Note that to calculate α, only the payoff of 2 is considered, and to get β, only the payoff of 1 is. This is because for an equilibrium the strategy of 1 has to be such that there is no incentive for 2 to change his strategy, and vice versa.

   Player 2's payoff at the equilibrium point is, then, 1/4, and player 1's is 2/3. An interesting aspect of the situation is that

player 1 can actually <u>guarantee</u> himself a payoff of 2/3 (by playing
the strategy α = 1/3), and player 2 can <u>guarantee</u> himself a payoff
of 1/4 (by playing the strategy β = 3/4). So at the equilibrium point,
the two players get what they can guarantee themselves: when each
player looks only at his own payoff and plays as if the game were strictly
competitive, each will receive his equilibrium payoff. But this behavior
does not generate the equilibrium <u>strategies</u>: if player 1 chooses
(1/3,2/3) and player 2 chooses (3/4,1/4), by changing to the strategy
(0,1) player 2 can improve his outcome. Conversely, use of the equilibrium strategies does not guarantee that the players will receive the
equilibrium payoffs: each player depends on the behavior of the other
to do so. In this case, then, the equilibrium point seems unconvincing
as a recommendation for a self-enforcing agreement, since each player
on his own can guarantee the exact amount yielded by this "agreement."

   <u>Exercise 1</u>: Find the equilibrium points and equilibrium payoffs
of the two-person game defined in the table below (each player has three
strategies).

	s_1^2	s_2^2	s_3^2
s_1^1	0,0	4,5	5,4
s_2^1	5,4	0,0	4,5
s_3^1	4,5	5,4	0,0

Exercise 2: Find the equilibrium points and equilibrium payoffs of the three-person game defined in the table below (each player has two strategies; player 3's strategies are to choose either of the matrices of payoffs).

	s_1^2	s_2^2
s_1^1	1,1,1	0,0,0
s_2^1	0,0,0	0,0,0

$$s_1^3$$

	s_1^2	s_2^2
s_1^1	0,0,0	0,0,0
s_2^1	0,0,0	2,2,2

$$s_2^3$$

Exercise 3: Find the equilibrium points and equilibrium payoffs of the two-person game defined in the table below.

	s_1^2	s_2^2
s_1^1	0,0	1,0
s_2^1	0,1	1,1

2.23 Theorem (Nash [1951]): The mixed extension of any game with finitely many strategies has an equilibrium point.

Proof: The proof requires the use of the following theorem.

Theorem (Brouwer's fixed point theorem): Let C be a compact convex subset of a Euclidean space. Let f be a continuous function from C into C. Then f has a fixed point; i.e. there is a point x in C for which $f(x) = x$.

Let $N = \{1,2,\ldots,n\}$ be the set of players and for each i in N let $S^i = \{1,2,\ldots,m^i\}$ be the set of pure strategies of player i. Let the payoff function of player i be $h^i(j^1,\ldots,j^k,\ldots,j^n) \in \mathbb{R}$ where j^k is the pure strategy chosen by player k $(1 \leq j^k \leq m^k)$. The corresponding mixed extension is defined by:

the player set $N = \{1,2,\ldots,n\}$,

the strategy space of player i: $X^i = \{(x_1, x_2, \ldots, x_{m^i}) \in E^{m^i}:$

$$x_j \geq 0 \text{ for all } j \text{ , and } \sum_{j=1}^{m^i} x_j = 1\} \text{ , and}$$

the payoff function of player i:

$$H^i(x^1,\ldots,x^k,\ldots,x^n) = \sum_{j^1=1}^{m^1} \sum_{j^2=1}^{m^2} \cdots \sum_{j^n=1}^{m^n} [x_{j^1}^1 \cdots x_{j^k}^k \cdots x_{j^n}^n h^i(j^1,\ldots,j^k,\ldots,j^n)] \quad .$$

X^i is by definition the simplex of dimension $m^i - 1$; we know that a simplex is convex and compact. We also know that the Cartesian product of compact convex sets is compact and convex, so that the Cartesian product of the strategy spaces $X = X^1 \times X^2 \times \ldots \times X^i \times \ldots X^n$ is compact and convex. Define the following function on X:

$$g_j^i(x) = \max (0, H^i(x|e_j^i) - H^i(x)) \qquad (\forall x \in X)(\forall i \in N)(\forall j \in S^i) \text{ ,}$$

where $e_j^i = (\overbrace{0,0,\ldots,0,1}^{j},0,\ldots,0)$ (i.e. e_j^i is the j-th unit vector in E^{m^i})

and $H^i(x|e_j^i)$ is the value of H^i when the mixed strategy of i in

x (x^i) is replaced by the pure strategy e_j^i. $H^i(x|e_j^i) - H^i(x)$ is,

then, the gain or loss accruing to player i as a consequence of his

move from x^i to e_j^i.

Define the function f: $X \to X$ by:

$$f_j^i(x) = \frac{x_j^i + g_j^i(x)}{1 + \sum\limits_{j=1}^{m^i} g_j^i(x)} \quad .$$

We see that $f(x) \in X$ and that $f(x)$ is continuous ($H^i(x)$ is conti-

nuous since it is a polynomial in x and $g(x)$ is continuous since it

is the maximum of two continuous functions). Hence, by Brouwer's theorem,

f has a fixed point; i.e. there is an x in X for which $f(x) = x$.

For this x, for all i and j,

(1) $\qquad x_j^i \sum\limits_{j=1}^{m^i} g_j^i(x) = g_j^i(x) \quad .$

Assertion: For all i in N, there is a j with $1 \leq j \leq m^i$

for which $x_j^i > 0$ and $g_j^i(x) = 0$.

Proof:

(2) $\qquad H^i(x) = \sum\limits_{j=1}^{m^i} x_j^i H^i(x|e_j^i) = \sum\limits_{x_j^i > 0} x_j^i H^i(x|e_j^i) \quad .$

If the assertion were false, then $g_j^i(x) > 0$ for all j such that $x_j^i > 0$, in which case

$$H^i(x|e_j^i) > H^i(x) \quad \text{for all } j \text{ such that } x_j^i > 0 \quad,$$

so that

$$(3) \qquad \sum_{x_j^i>0} x_j^i H^i(x|e_j^i) > \sum_{x_j^i>0} x_j^i H^i(x) = H^i(x) \sum_{x_j^i>0} x_j^i = H^i(x) \quad.$$

(2) and (3) involve a contradiction, so that the assertion is proved.

Applying the assertion, (1) leads to $\sum_{j=1}^{m^i} g_j^i(x) = 0$ for all i, and since $g_j^i \geq 0$, $g_j^i(x) = 0$ for all i and j. Hence $H^i(x|e_j^i) \leq H^i(x)$, and for every mixed strategy $y^i \in X^i$,

$$H^i(x|y^i) = \sum_{j=1}^{m^i} y_i H(x|e_j^i) \leq \sum_{j=1}^{m^i} y_j H^i(x) = H^i(x) \sum_{j=1}^{m^i} y_j = H^i(x) \quad,$$

since $\sum_{j=1}^{m^i} y_j = 1$. We conclude that x is an equilibrium point, which establishes the theorem.

Chapter 3: The Shapley Value

In this and subsequent chapters, we turn to the theory of "cooperative games," where the focus of interest is the way in which the players bargain together over the division of the available payoff, rather than the way this payoff can be attained by the use of certain strategies.

3.1 Definition: A game in coalitional form consists of

  1) a set N (the players), and

  2) a function $v: 2^N \to \mathbb{R}$ such that $v(\emptyset) = 0$. ($2^N = \{S: S \subset N\}$).

A subset of N is called a coalition; $v(S)$ is called the worth of the coalition S.

3.2 Agreement: If $\{i_1, i_2, \ldots, i_J\}$ is a set of players, we will some-times write $v(i_1 i_2 \ldots i_J)$ rather than $v(\{i_1, i_2, \ldots, i_J\})$ for the worth of $\{i_1, i_2, \ldots, i_J\}$.

3.3 Example (2-person bargaining game):

  $N = \{1,2\}$, $v(N) = 1$, $v(1) = v(2) = 0$.

3.4 Example (Market for a perfectly divisible good with one buyer and two sellers):

  $N = \{1,2,3\}$, $v(N) = v(12) = v(13) = 1$, $v(23) = v(1) = v(2) = v(3) = 0$

3.5 Example (Pure bargaining game with n players, or unanimity game with n players):

  $v(N) = 1$, $v(S) = 0$ for $S \neq N$.

3.6 Example (3-person majority game):

  $N = \{1,2,3\}$, $v(N) = v(12) = v(13) = v(23) = 1$, $v(1) = v(2) = v(3) = 0$

3.7   Example (Weighted majority game):

$$N = \{1,2,3,4\} \quad v(S) = \begin{cases} 1 & \text{if } \sum_{i \in S} w^i \geq 3 \\ 0 & \text{if } \sum_{i \in S} w^i \leq 2 \end{cases} \quad ,$$

with $w^1 = 2$ and $w^i = 1$ for $i = 2,3,4$.

(w^i is the "weight" of player i.)

3.8   Definition: Let N be the set of players. An n-person weighted majority game with weights $\{w^i\}_{i \in N}$ and quota q is defined by

$$v(S) = \begin{cases} 1 & \text{if } \sum_{i \in S} w^i \geq q \\ 0 & \text{if } \sum_{i \in S} w^i < q \end{cases} \quad .$$

3.9   Definition: v is monotonic if $S \supset T$ implies $v(S) \geq v(T)$.
(Note that this does not mean that $|S| \geq |T|$ implies $v(S) \geq v(T)$
(where $|S|$ is the cardinality of S).) v is superadditive if
$S \cap T = \emptyset$ implies $v(S \cup T) \geq v(S) + v(T)$.

Unless specifically stated, it will not be assumed that v is
monotonic or superadditive.

3.10   Definition: A game is 0-normalized if $v(i) = 0$ for all i in
N; it is 0-1 normalized if it is 0-normalized and $v(N) = 1$.

3.11   Definition: i and j, elements of N, are substitutes in v
if for all S containing neither i nor j, $v(S \cup \{i\}) = v(S \cup \{j\})$.

3.12   Definition: An element i of N is called a null player if
$v(S \cup \{i\}) = v(S)$ for all $S \subseteq N$.

3.13   Definition: E^N is the Euclidean space whose dimension is the
cardinality of N, and whose coordinates are indexed by the members
of N themselves.

We now introduce the solution concept studied in this chapter.

3.14   Definition: Let $N = \{1,2,\ldots,n\}$ and let G^N be the set of
all games whose player set is N. A Shapley value or value on N is a
function ϕ: $G^N \to E^N$ satisfying the following conditions:

1. (Symmetry condition): if i and j are substitutes in v,
    then $(\phi v)_i = (\phi v)_j$.

2. (Null player condition): if i is a null player, then
    $(\phi v)_i = 0$.

3. (Efficiency condition): $\sum_{i=1}^{n} (\phi v)_i = v(N)$.

4. (Additivity condition): $(\phi(v + w))_i = (\phi v)_i + (\phi w)_i$.

3.15   Remark: $(\phi v)_i$, the i-th coordinate of the image vector $\phi(v)$
(sometimes denoted ϕv) is interpreted as the "power" of player i in
the game v, or what it is worth to i to participate in the game v
(in brief, v's "value" for i).

3.16   Remark: Conditions 1, 2, and 4 are weak restrictions which are
easy to accept as "reasonable," while 3 is much stronger (to require an
efficient outcome in game situations is as strong an assumption as requir-
ing it in a traditional economic problem).

3.17   Theorem (Shapley [1953a]): There exists a unique value on G^N for every N.

Proof: First we prove uniqueness. Let ϕ be a value on G^N. Define for each coalition $T \subseteq N$ with $T \neq \emptyset$, a game v_T by

$$v_T(S) = \begin{cases} 1 & \text{if } T \subseteq S \\ 0 & \text{otherwise} \end{cases} .$$

Note that for any real α, members of $N \backslash T$ are null players in αv_T, and members of T are substitutes for each other in αv_T. Hence by the null player condition, $\phi(\alpha v_T)_i = 0$ when $i \notin T$, and by the symmetry condition $\phi(\alpha v_T)_i = \phi(\alpha v_T)_j$ when $i, j \in T$. Hence, by the efficiency condition $\sum_{i \in N} \phi(\alpha v_T)_i = (\alpha v_T)(N) = \alpha v_T(N) = \alpha$. Thus $\alpha = \sum_{i \in T} \phi(\alpha v_T)_i = |T| \phi(\alpha v_T)_i$ for any $i \in T$. Hence,

$$\phi(\alpha v_T)_i = \begin{cases} \dfrac{\alpha}{|T|} & \text{for } i \in T \\ 0 & \text{for } i \notin T \end{cases} .$$

Now, G^N is a Euclidean space of dimension $2^{|N|} - 1$ and there are $2^{|N|} - 1$ games v_T. We know $\phi(\alpha v_T)$ for all α and T, so by additivity we know $\phi(\sum_{i=1}^{k} \alpha_i v_{T_i})$ for all linear combinations $\sum_{i=1}^{k} \alpha_i v_{T_i}$ of the v_T's. Hence if we prove that the v_T's are linearly independent, we will have shown uniqueness. Suppose they are not; then we may write

$$v_T = \sum_{i=1}^{j} \beta_i v_{T_i} \quad ,$$

where $|T| \leq |T_i|$ for all i and all T_i's are different from each other and from T. Then

$$1 = v_T(T) = \sum_{i=1}^{j} \beta_i v_{T_i}(T) = \sum_{i=1}^{j} \beta_i \cdot 0 = 0 \quad,$$

a contradiction. We therefore conclude that the v_T's are indeed linearly independent, which completes the uniqueness proof.

For the existence proof, suppose that the players in N are ordered, and suppose that according to this order, each player gets his marginal incremental worth to the coalition formed by the players preceding him. That is, the i-th player gets

$$v(1,2,3,\ldots,i-1,i) - v(1,2,3,\ldots,i-1) \quad,$$

where $1,\ldots,i-1$ denotes the players before i in the order under consideration. The function on N thus obtained does not always satisfy the conditions of the Shapley value. But if we take all possible orders of the players and average the corresponding marginal contributions, this average turns out to satisfy all the conditions of the Shapley value. Thus, a null player has zero incremental worth in all orders, and the symmetry of the set of all orderings ensures that the symmetry condition is satisfied. The efficiency condition may also be verified, and the additivity follows from

$$(v + w)(1,2,\ldots,i-1,i) - (v + w)(1,2,\ldots,i-1)$$
$$= [v(1,2,\ldots,i-1,i) - v(1,2,\ldots,i-1)] + [w(1,2,\ldots,i-1,i)$$
$$- w(1,2,\ldots,i-1)]$$

for any two games v and w and any order. This establishes the existence of a Shapley value, so that the proof of the theorem is now complete. The above argument also establishes the following.

3.18   Theorem (Shapley [1953a]): $(\phi v)_i = (1/|N|!)\sum_R [v(S_i \cup \{i\}) - v(S_i)]$ where R runs over all $|N|!$ different orders on N, and S_i is the set of players preceding i in the order R.

We will now compute the Shapley value for some simple games.

3.19   Example: 2-person bargaining game. One has

$$N = \{1,2\} \quad v(12) = 1 \quad , \quad v(1) = v(2) = 0 \quad ,$$

so the formula gives:

$$(\phi v)_1 = (\phi v)_2 = \frac{1}{2} \quad .$$

3.20   Example: 3-person majority game. One has

$$N = \{1,2,3\} \quad v(1) = v(2) = v(3) = 0 \quad ,$$
$$v(12) = v(23) = v(31) = v(123) = 1 \quad ,$$

so the formula gives:

$$(\phi v)_1 = (\phi v)_2 = (\phi v)_3 = \frac{1}{3} \quad .$$

In both these examples one can also deduce the value directly from the symmetry and efficiency conditions.

3.21 <u>Example</u>: Market with two sellers and one buyer. Here

$$N = \{1,2,3\} \quad , \quad v(123) = v(12) = v(13) = 1 \quad ,$$

$$\text{and} \quad v(S) = 0 \quad \text{for all other} \quad S \subset N \quad .$$

In order to compute the Shapley value for this game, we first notice that there are $3! = 6$ orderings of the 3 players. Since this game is a <u>simple game</u> (i.e. the worth of every coalition is either 0 or 1), the following definition is useful: player i is a <u>key player</u> with respect to the coalition S if $v(S) = 0$ and $v(S \cup \{i\}) = 1$. The Shapley value for a player i is his average incremental worth, so we obtain it by computing the proportion of orderings in which player i is a key player with respect to the set of players which precedes him in the ordering. The six orderings are:

$$\{1,2,3\}, \{1,3,2\}, \{2,1,3\}, \{2,3,1\}, \{3,1,2\}, \text{ and } \{3,2,1\}.$$

Player 1 is key in $\{2,1,3\}$, $\{2,3,1\}$, $\{3,1,2\}$, and $\{3,2,1\}$. So

$$(\phi v)_1 = \frac{4}{6} = \frac{2}{3} \quad .$$

Since 2 and 3 are substitutes $(\phi v)_2 = (\phi v)_3$. The efficiency condition is $\sum_{i=1}^{3} (\phi v)_i = v(123) = 1$, so

$$(\phi v)_2 = (\phi v)_3 = \frac{1}{6} \quad ,$$

so that

$$\phi v = (\frac{2}{3},\frac{1}{6},\frac{1}{6}) \quad .$$

This example illustrates the fact that the Shapley value gives a measure
of the power of the players in a situation free of any institutions.
Thus, one might think that if the available payoff above were distributed
according to the players' strengths, the outcome would be $(1/2,1/4,1/4)$,
since the two sellers can form a cartel which will put them on an equal
footing with the buyer. The Shapley value, however, reflects the fact
that the buyer is actually in a stronger position since each of the
sellers may be willing to deal with him separately.

3.22 Examples: Weighted majority games. The Shapley value gives
interesting insights into some multi-party political situations. For
instance, the political arena in Israel is characterized by the existence of a large party (the Labor Party) which counts for approximately
1/3 of the votes, whereas until several years ago the remaining votes
were split among many relatively small parties. In spite of the fact
that it controlled only 1/3 of the votes, however, the Labor Party has,
since the creation of the state, always held all four major ministries
(Prime Minister, Finance, Foreign, Defense).

To try to gain some insight into this situation, let us compute
the Shapley value for a weighted majority game (N,v) with quota
$q = 1/2$ and a vector of weights $w = (1/3,2/9,2/9,2/9)$. We get
$\phi v = (1/2,1/6,1/6,1/6)$. This result provides some understanding of the
situation in Israel: although it has only 1/3 of the votes, the Labor
Party has half the "power" within parliament.

Consider next a situation in which there are 100 parties: one large party has 1/3 of the votes, and the remaining 99 parties share the other 2/3 equally. The large party is a key player in all orderings in which there are more than 1/4 and less than 3/4 of the 99 players before him. So he is key in half of the orderings, so that again 1/3 of the votes gives the large party 1/2 of the power: $(\phi v)_1 = 1/2$.

Now consider a situation in which there are two large parties, each with 1/3 of the votes, and 3 small ones with 1/9 of the votes each; i.e.

$$w = (\frac{1}{3}, \frac{1}{3}, \frac{1}{9}, \frac{1}{9}, \frac{1}{9}) \quad .$$

We will compute the Shapley value for the corresponding weighted majority game with $q = 1/2$. Let the two large players be denoted by x and y. For each order on the small players, one can characterize the order on all the players by a pair (a,b), where a (resp. b) is the number of small players after which x (resp. y) appears. Corresponding to a pair (α,α) there are two orders of all the players--one where x precedes y, and one where the reverse is true; corresponding to every other pair there is just one order. Hence possible orders are illustrated in the diagram below: for example, the point A corresponds to the order (p_1, y, p_2, x, p_3) (where (p_1, p_2, p_3) is the ordering of the small players); each position on the diagonal corresponds to two possible orders of all players. So for every ordering of the small players there are 20 possible orderings of all the players, and x is a key player in the six positions which

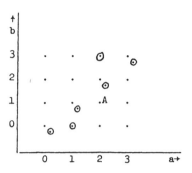

are circled. So, since the order of the small players is irrelevant at present, the value of x is 6/20 = 3/10. By symmetry the value of y is also 3/10, and by symmetry and efficiency the value of the game is

$$\phi v = (\frac{3}{10},\frac{3}{10},\frac{2}{15},\frac{2}{15},\frac{2}{15}) \quad .$$

So in this case, the Shapley value imputes to each of the large players a share of the power smaller than his share of the votes.

Let us consider now a more general case in which there are two large parties (each with 1/3 of the votes) and n - 2 small ones of equal size. We are interested in the Shapley value of this game for n arbitrarily large. The characterization of orderings used above can be modified by letting a and b be the proportions of the small players after which x and y respectively appear. The diagram below then illustrates the situation, the shaded area corresponding to those order-ings in which x is a key player. The value of each of the large parties is, then, approximately 1/4 when n is large; for n = 5, it was 3/10. If there is a large number of small parties it will, then, be better for

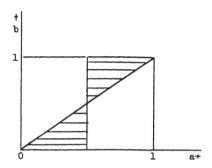

them not to get together in larger groups. The intuitive rationale is
as follows. Whatever the number of small parties, each large party
does not need all their votes to form a majority, but if there are few
small parties the large ones will have no choice but to bargain over
large blocks of votes. If there is a large number of small parties,
the large parties can bargain for just the number of votes they need,
and can consequently offer more per vote: the small parties will then
actually be more powerful.

This result may account for another aspect of the political scene
in Israel: the fact that the relatively small religious parties have
not gotten together, but have remained independent, in the presence of
the two large parties on the right and on the left.

Chapter 4: The Core

4.1   Definition: A payoff vector is a member of E^N (the Euclidean
$|N|$-dimensional space whose coordinates are indexed by the members of N).

4.2   Definition: A payoff vector x is called individually rational
(in the game (N,v)) if $x^i \geq v(i)$ for all players $i \in N$.

4.3　Definition: A payoff vector x is called group rational (or efficient) if $\sum_{i \in N} x^i = v(N)$.

4.4　Remark: If $\sum_{i \in N} x^i < v(N)$, then all players could improve their payoff by forming the coalition N; hence x is inefficient. If v is superadditive, then for any partition $\{S_1, \ldots, S_k\}$ of the players (i.e. $\bigcup_{i=1}^{k} S_i = N$ and $S_i \cap S_j = \emptyset$ for all $i \neq j$), we have $v(N) \geq \sum_{i=1}^{k} v(S_i)$; therefore there is no way for the players to obtain a total payoff greater than $v(N)$. Hence under the assumption of super-additivity, it is to be expected that payoff vectors that actually occur will be group rational. However, superadditivity will not be assumed here unless specifically stated.

4.5　Definition: An imputation is a payoff vector that is indivi-dually and group rational.

4.6　Definition: The core of the game (N, v) is the set of all imputa-tions x such that $v(S) \leq \sum_{i \in S} x^i$ for all $S \subset N$.

4.7　Example: Two-person bargaining game. We have $N = \{1, 2\}$, $v(N) = 1$, and $v(1) = v(2) = 0$. Then (x^1, x^2) is in the core if and only if

$$x^1 \geq 0 \quad, \quad x^2 \geq 0 \quad, \quad \text{and} \quad x^1 + x^2 = 1 \quad.$$

So the core is the set of all imputations, as shown in the diagram.

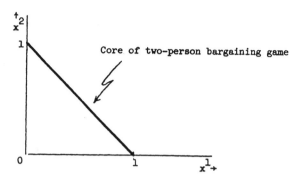

Core of two-person bargaining game

4.8   **Example**: Three-person bargaining game. In this game $N = \{1,2,3\}$,
$v(N) = 1$ and $v(S) = 0$ for all other $S \subset N$. So (x^1, x^2, x^3) is in the
core if and only if

$$x^1 + x^2 + x^3 = v(N) = 1 \quad , \quad x^i \geq v(i) = 0 \quad \text{for all} \quad i \in N \quad , \quad \text{and}$$

$$\sum_{i \in S} x^i \geq v(S) = 0 \quad \text{for all} \quad S \subset N \quad , \quad S \neq N \quad .$$

The core is therefore the set of all imputations once again; it is shown
in the diagram below.

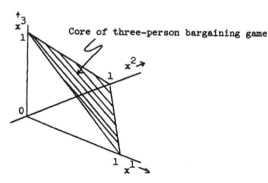

Core of three-person bargaining game

4.9   Example: Market with 2 sellers and a buyer. In this game
$N = \{1,2,3\}$, $v(123) = v(12) = v(13) = 1$, and $v(S) = 0$ for all other
$S \subset N$. So x is in the core if and only if

$$x^1 + x^2 + x^3 = 1 \quad , \quad x^1 + x^2 \geq 1 \quad , \quad x^2 + x^3 \geq 1 \quad ,$$

$$x^1 \geq 0 \quad , \quad x^2 \geq 0 \quad , \quad \text{and} \quad x^3 \geq 0 \quad .$$

Hence the core is $\{(1,0,0)\}$.

4.10   Remark: Note that the core in the example above $(\{(1,0,0)\})$
differs considerably from the Shapley value of the game considered there
(which is $(2/3,1/6,1/6)$). One can interpret the zero payoff to players
2 and 3 in the core allocation as the result of cutthroat competition
between them.

4.11   Example: 3-person majority game. Here $N = \{1,2,3\}$,
$v(123) = v(12) = v(13) = v(23) = 1$, and $v(i) = 0$ for all $i \in N$.
For x to be in the core, we need $x^1 + x^2 + x^3 = 1$, $x^i \geq 0$ for all
$i \in N$, $x^1 + x^2 \geq 1$, $x^1 + x^3 \geq 1$, and $x^2 + x^3 \geq 1$. There exists no x
satisfying these conditions, so the core is empty.

We now wish to study conditions on v which will ensure that the
core of (N,v) be non-empty. Consider first a 0-1 normalized 3-person
game. Let us suppose that the core is non-empty, i.e. there exists an
imputation $x = (x^1,x^2,x^3)$ such that

$$x^1 + x^2 \geq v(12) \quad ,$$

$$x^1 + x^3 \geq v(13) \quad ,$$

and

$$x^2 + x^3 \geq v(23) \quad .$$

In this case we have

$$2(x^1 + x^2 + x^3) \geq v(12) + v(13) + v(23) \quad ,$$

or

$$v(N) = 1 \geq \frac{[v(12) + v(13) + v(23)]}{2} \quad ,$$

and

$$v(12) \leq 1 \quad , \quad v(13) \leq 1 \quad , \quad v(23) \leq 1 \quad .$$

So a necessary condition for a 0-1 normalized 3-person game to have a non-empty core is that $1 \geq [v(12) + v(13) + v(23)]/2$ and $v(ij) \leq 1$ for all $\{i,j\} \subset N$.

Exercise 4: Prove that the condition $1 \leq [v(12) + v(13) + v(23)]/2$ and $v(ij) \leq 1$ for all $\{i,j\} \subset N$ is also a sufficient condition for the 0-1 normalized 3-person game (N,v) to have a non-empty core.

Let us now consider the conditions under which a general game (N,v) has a non-empty core.

4.12   Definition: Let $S \subset N$. The <u>characteristic vector of</u> S is
the element χ_S of E^N defined by

$$\chi_S^i = \begin{cases} 1 & \text{if } i \in S \\ 0 & \text{otherwise} \end{cases}$$

4.13   Definition: A family S of coalitions is called <u>balanced</u> if
there exists a sequence of non-negative numbers $\{\delta_S\}_{S \in S}$ such that

$$\sum_{S \in S} \delta_S \chi_S = \chi_N \quad .$$

$\{\delta_S\}_{S \in S}$ are called <u>balancing weights</u> for S .

A natural interpretation of this definition is the following.
Each player is endowed with one unit of time that he allocates among the
coalitions S in S ; δ_S is the fraction of his time that each member
of S allocates to the coalition S ; the condition $\sum_{S \in S} \delta_S \chi_S = \chi_N$ is a
feasibility condition (for every individual the sum of the amounts of
his time he spends with each coalition must equal exactly the amount of
time he is endowed with).

4.14   Theorem (Bondareva [1962], [1963], and Shapley [1967]): <u>A necessary</u>
<u>and sufficient condition for the core of</u> (N,v) <u>to be non-empty is that for</u>
<u>all balanced families</u> S <u>and corresponding balancing weights</u> $\{\delta_S\}_{S \in S}$,
<u>we have</u> $\sum_{S \in S} \delta_S v(S) \leq v(N)$.

Proof: We will assume that v is 0-1 normalized; the extension
to the general case is left to the reader.

1. <u>The condition is necessary</u>.

Let x be in the core. Then $\sum_{i \in N} x^i = v(N)$ and $\sum_{i \in S} x^i \geq v(S)$ for all $S \subset N$. Let S be a balanced family with weights $\{\delta_S\}$. Then

$$\delta_S \sum_{i \in S} x^i \geq \delta_S v(S) \quad ,$$

so

$$\sum_{S \in S} \sum_{i \in S} \delta_S x^i \geq \sum_{S \in S} \delta_S v(S) \quad .$$

Since we are dealing with a finite sum we can reverse the double summation sign:

$$\sum_{S \in S} \sum_{i \in S} \delta_S x^i = \sum_{i \in N} \sum_{\substack{S \in S \\ S \ni i}} \delta_S x^i = \sum_{i \in N} x^i \sum_{\substack{S \in S \\ S \ni i}} \delta_S = \sum_{i \in N} x^i = v(N) \quad .$$

Hence

$$v(N) \geq \sum_{S \in S} \delta_S v(S) \quad .$$

This establishes necessity.

2. <u>The condition is sufficient</u>.

Assume $v(N) \geq \sum_{S \in S} \delta_S v(S)$ for all balanced families S and corresponding weights. Define a 2-person 0-sum game as follows. Player I chooses a player i in the game (N,v). Player II chooses a coalition S in the game (N,v), such that $v(S) > 0$. The payoff to Player I is:

$$h(i,S) = \begin{cases} \dfrac{1}{v(S)} & \text{if } i \in S \\ 0 & \text{otherwise} \end{cases} \quad .$$

Assertion: In order to prove that the condition is sufficient, it is enough to prove that the minimax value of this 2-person game is greater than or equal to 1.

Proof: If the minimax value is greater than or equal to 1, there is a mixed strategy x of Player I that yields at least 1, no matter which pure strategy S is chosen by Player II. That is,

$$1 \leq \sum_{i \in N} x^i h(i,S) = \frac{1}{v(S)} \sum_{i \in S} x^i$$

for all $S \subset N$ with $v(S) > 0$. Hence

$$v(S) \leq \sum_{i \in S} x^i$$

for all $S \subset N$ such that $v(S) > 0$. When $v(S) = 0$, the inequality holds since $x^i \geq 0$ for all i. Hence

$$v(S) \leq \sum_{i \in S} x^i$$

for all $S \subset N$. Together with the condition $\sum_{i \in N} x^i = 1$, this means that x is in the core of the game (N,v), so that that core is non-empty. This establishes the assertion.

So we must now prove that the 2-person 0-sum game has a minimax value greater than or equal to 1. Suppose contrariwise that the minimax value is less than 1; let it be $0 < \xi < 1$ (notice that $\xi > 0$, since if Player I chooses a strictly positive probability for every player, he will be guaranteed a positive payoff). There is then a mixed strategy for Player II that guarantees that the payoff to I will at most be ξ. Let this mixed strategy assign probability $\theta_S > 0$ to each coalition in a family S with $v(S) > 0$ for all $S \in S$. For each i, we have

$$\xi \geq \sum_{\substack{S \in S \\ S \ni i}} \theta_S h(i,S) = \sum_{\substack{S \in S \\ S \ni i}} \frac{\theta_S}{v(S)} \quad ,$$

so

$$1 \geq \sum_{\substack{S \in S \\ S \ni i}} \frac{\theta_S}{\xi v(S)} \quad .$$

Let us define $\delta_S = \theta_S/\xi v(S)$ for all $S \in S$. Then we have

$$1 \geq \sum_{\substack{S \in S \\ S \ni i}} \delta_S \quad .$$

In order to construct a balanced family of coalitions, define

$$\delta_i = 1 - \sum_{\substack{S \in \mathcal{S} \\ S \ni i}} \delta_S \quad .$$

Consider the collection $\mathcal{T}$ consisting of $\mathcal{S}$ and all singletons $\{i\}$. Then for all i,

$$\sum_{\substack{S \in \mathcal{T} \\ S \ni i}} \delta_S = \sum_{\substack{S \in \mathcal{S} \\ S \ni i}} \delta_S + \delta_i = 1 \quad ,$$

so $\mathcal{T}$ is a balanced family with balancing weights $\{\delta_S\}$. Hence by assumption

$$\sum_{S \in \mathcal{T}} \delta_S v(S) \leq v(N) \quad ,$$

so that, since $v(\{i\}) = 0$ for all i,

$$\sum_{S \in \mathcal{S}} \delta_S v(S) \leq v(N) \quad .$$

So

$$\sum_{S \in \mathcal{S}} \frac{\theta_S}{\xi} \leq v(N) = 1 \quad ,$$

or

$$\sum_{S \in \mathcal{S}} \theta_S \leq \xi < 1$$

(we have supposed $\xi < 1$). But this result contradicts the fact that $\{\theta_S\}_{S \in \mathcal{S}}$ is a strategy for Player II: we need $\sum_{S \in \mathcal{S}} \theta_S = 1$. Hence the minimax value is greater than or equal to 1, which, using the above assertion, establishes sufficiency.

The following leads up to an exercise in the use of the Bondareva-Shapley theorem.

4.15   Definition: S is a winning coalition in a simple game if $v(S) = 1$; a veto player in such a game is a player who is a member of every winning coalition.

Exercise 5: Prove that a 0-1 normalized weighted majority game has a non-empty core if and only if there is at least one veto player.

Exercise 6: Find the core of a 0-1 normalized weighted majority with $p \geq 1$ veto players.

We may sum up some basic features of the Shapley value and the core as follows:

The Shapley value of a game is a single payoff vector. It is always group rational; in superadditive games it is individually rational, but this is not necessarily so in general.

The core is a set of payoff vectors. It is a subset of the set of imputations. It may be empty, and even when it is not the Shapley value may not be a member of it.

Intuitively the Shapley value represents a "reasonable compromise", whereas the core represents a set of payoff vectors which are in a certain sense "stable". There is no general relationship

between the two, though for certain classes of games (not considered
in these Lectures) a close relationship can be established.

Chapter 5: Market Games

Let us now consider an economic application of the concepts we
have developed. The situation we will describe is that of a "market
game". In a market game, there is one consumption good, ℓ production
goods and n players. Each player i has a <u>production function</u>
$u^i(x_1, x_2, \ldots, x_\ell)$, defined for all $x_j \geq 0$ and with values in $\mathbb{R}$. The
quantity $u^i(x_1, x_2, \ldots, x_\ell)$ represents the amount of the single consumption good that i can produce from inputs $x_1, x_2, \ldots, x_\ell$. Each player
i also has an <u>endowment</u> $(a_1^i, a_2^i, \ldots, a_\ell^i)$ of production goods.
Each coalition produces as much of the consumption good as possible so
that

$$(1) \qquad v(S) = \max \{ \sum_{i \in S} u^i(x^i): \sum_{i \in S} x^i = \sum_{i \in S} a^i \text{ and } x^i \geq 0 \text{ for all } i\}$$

where $x^i = (x_1^i, x_2^i, \ldots, x_\ell^i)$ and a^i is similarly defined.

5.1   <u>Remark</u>: If the u^i's are continuous, then the above maximum is
attained.

5.2   <u>Definition</u>: A function u is called <u>concave</u> if its domain is
convex and for x and y in the domain of u and all α in $[0,1]$,

$$u(\alpha x + (1 - \alpha)y) \geq \alpha u(x) + (1 - \alpha)u(y) \quad .$$

5.3   Definition: Assume that the u^i's are concave and continuous. Then the game (N,v) defined by (1) is called a market game.

5.4   Proposition (Shapley and Shubik [1969]): Every market game has a non-empty core.

Proof: We will use the Bondareva-Shapley theorem. Let S be a balanced collection of coalitions with corresponding weights $\{\delta_S\}_{S \in \mathcal{S}}$. We must prove that

$$\sum_{S \in \mathcal{S}} \delta_S v(S) \leq v(N) \quad .$$

Let $v(S) = \sum_{i \in S} u^i(x_S^i)$ where x_S^i is the point of the set $\{y^i \in E_+^\ell : \sum_{i \in S} y^i = \sum_{i \in S} a^i\}$ at which the function $\sum_{i \in S} u^i(y^i)$ attains its maximum. Define

$$x^i = \sum_{\substack{S \in \mathcal{S} \\ S \ni i}} \delta_S x_S^i \quad .$$

(One can think of player i spending a fraction δ_S of his time in coalition S; x^i is then his total input vector.) We can then prove that $(x^i)_{i \in N}$ is a feasible allocation for N:

$$\sum_{i \in N} x^i = \sum_{i \in N} \sum_{\substack{S \in \mathcal{S} \\ S \ni i}} \delta_S x_S^i = \sum_{S \in \mathcal{S}} \sum_{i \in S} \delta_S x_S^i = \sum_{S \in \mathcal{S}} \delta_S \sum_{i \in S} x_S^i$$

$$= \sum_{S \in \mathcal{S}} \delta_S \sum_{i \in S} a^i = \sum_{S \in \mathcal{S}} \sum_{i \in S} \delta_S a^i = \sum_{i \in N} \sum_{\substack{S \in \mathcal{S} \\ S \ni i}} \delta_S a^i = \sum_{i \in N} a^i \sum_{\substack{S \in \mathcal{S} \\ S \ni i}} \delta_S = \sum_{i \in N} a^i$$

(since $\sum\limits_{\substack{S \in S \\ S \ni i}} \delta_S = 1$ for all i). Hence

$$\sum_{i \in N} x^i = \sum_{i \in N} a^i \quad .$$

Moreover, the x^i's are non-negative since they are averages of non-negative numbers with positive weights. So $(x^i)_{i \in N}$ is a feasible allocation for N. Hence by the definition of $v(N)$

$$v(N) \geq \sum_{i \in N} u^i(x^i) \quad .$$

Since u^i is a concave function

$$u^i(x^i) \geq \sum_{\substack{S \in S \\ S \ni i}} \delta_S u^i(x_S^i) \quad .$$

Hence

$$v(N) \geq \sum_{i \in N} \sum_{\substack{S \in S \\ S \ni i}} \delta_S u^i(x_S^i)$$

$$= \sum_{S \in S} \delta_S \sum_{i \in S} u^i(x_S^i) = \sum_{S \in S} \delta_S v(S) \quad .$$

So

$$\sum_{S \in S} \delta_S v(S) \leq v(N) \quad .$$

So by the Bondareva-Shapley theorem the core is non-empty, which establishes the proposition.

The converse of Proposition 5.4 is false—not every game with a non-empty core is a market game. For example, the four-person game defined by $v(1234) = 2$, $v(S) = 1$ if $|S| = 2$ or 3, and $v(S) = 0$ if $|S| = 0$ or 1, is not a market game, although $(1/2,1/2,1/2,1/2)$ is in the core.

5.5 Definition: Let (N,v) be a game, and $T \subseteq N$. The subgame (T,v_T) defined by T is the game whose player set is T and whose worth function is defined by $v_T(S) = v(S)$ for all $S \subseteq T$.

Obviously every subgame of a market game is itself a market game, and so from Proposition 5.4 we obtain

5.6 Corollary: Every subgame of a market game has a non-empty core.

The 4-person game defined above has a subgame (defined by $T = \{1,2,3\}$, say) with an empty core. This raises the question whether every game, all of whose subgames have non-empty cores, is a market game. This is indeed the case; we have

5.7 Theorem (Shapley and Shubik [1969]): A necessary and sufficient condition for a game (N,v) to be a market game is that it and all of its subgames have non-empty cores.

Proof: We have already proved that the condition is necessary. To prove that the condition is sufficient, we consider a game (N,v) such that it and all of its subgames have non-empty cores. We will construct a market game such that its value function is precisely v.

Define a market by $\ell = n$; good i is the labor time of player i. The endowment of player i is defined by the i-th unit vector of the Euclidean space E^n: $a^i = \underbrace{(0,0,0,\ldots,0,1,0,\ldots,0)}_{i}$ (i.e. each player is endowed with one unit of his own labor time). The players have the same production functions, defined by

$$u^i(x) = u(x) = \max_{\{\alpha_T\}} \{ \sum_{T \subset N} \alpha_T v(T): \alpha_T \geq 0 \text{ and } \sum_{T \subset N} \alpha_T \chi_T = x \} \ .$$

Let

$$w(S) = \max \{ \sum_{i \in S} u(x^i): \sum_{i \in S} x^i = \chi_S \} \ .$$

(N,w) is, then, a market game; we will show that $w(S) = v(S)$ for all $S \subset N$. By the definition of $w(S)$,

$$w(S) \leq u(\chi_S)$$

$$= \max_{\{\alpha_T\}} \{ \sum_{T \subset N} \alpha_T v(T): \alpha_T \geq 0 \text{ and } \sum_{T \subset N} \alpha_T \chi_T = \chi_S \}$$

$$\geq v(S)$$

(taking $\alpha_S = 1$, and $\alpha_T = 0$ for all $T \neq S$). So we have proved that $w(S) \geq v(S)$. In order to prove the reverse inequality, we are going to use the hypothesis that every subgame has a non-empty core. We want to prove that $w(S) \leq v(S)$. We will first prove that $w(S) \leq u(\chi_S)$, and then that $u(\chi_S) \leq v(S)$.

Let the maximum in the definition of w by attained at x_S, so that

$$w(S) = \sum_{i \in S} u(x_S^i) \quad .$$

We will show that $\sum_{i \in S} u(x_S^i) \leq u(\sum_{i \in S} x_S^i)$ for all $S \subset N$, i.e. that u is superadditive.

Assertion: u is homogeneous of degree 1, i.e. for all $\alpha \geq 0$, $u(\alpha x) = \alpha u(x)$.

Proof:

$$u(\alpha x) = \max_{\{\alpha_T\}} \{ \sum_{T \subset N} \alpha_T v(T) : \alpha_T \geq 0 \text{ and } \sum_{T \subset N} \alpha_T x_T = \alpha x \}$$

$$= \max_{\{\alpha_T\}} \{ \alpha \sum_{T \subset N} \frac{\alpha_T}{\alpha} v(T) : \alpha_T \geq 0 \text{ and } \sum_{T \subset N} \frac{\alpha_T}{\alpha} x_T = x \}$$

$$= \alpha \max_{\{\beta_T\}} \{ \sum_{T \subset N} \beta_T v(T) : \beta_T \geq 0 \text{ and } \sum_{T \subset N} \beta_T x_S = x \}$$

$$= \alpha u(x)$$

(with $\beta_T = \alpha_T/\alpha$).

Assertion: u is a concave function, i.e. for all $1 \geq \alpha \geq 0$, $u[\alpha x + (1 - \alpha)y] \geq \alpha u(x) + (1 - \alpha)u(y)$.

Proof: Let

$$u(x) = \sum \alpha_T v(T) \qquad \text{with} \quad \sum \alpha_T x_T = x$$

and

$$u(y) = \sum \beta_T v(T) \qquad \text{with} \quad \sum \beta_T x_T = y \quad.$$

By definition

$$u[\alpha x + (1 - \alpha)y] \geq \sum [\alpha\alpha_T + (1 - \alpha)\beta_T]v(T) \quad,$$

since

$$\sum [\alpha\alpha_T + (1 - \alpha)\beta_T]x_T = \alpha\sum\alpha_T x_T + (1 - \alpha)\sum\beta_T x_T$$

$$= \alpha x + (1 - \alpha)y \quad.$$

Hence

$$u[\alpha x + (1 - \alpha)y] \geq \alpha\sum\alpha_T v(T) + (1 - \alpha)\sum\beta_T v(T)$$

$$= \alpha u(x) + (1 - \alpha)u(y) \quad.$$

Exercise 7: Prove that the following is true for every concave function $f: E^n \to \mathbb{R}$ and for all $m \geq 1$:

$$\forall \alpha \in E_+^m \quad, \quad \sum_{i=1}^m \alpha_i = 1 \Rightarrow f\left(\sum_{i=1}^m \alpha_i x_i\right) \geq \sum_{i=1}^m \alpha_i f(x_i) \quad.$$

We deduce the superadditivity of u from the two assertions above:

$$\sum_{i\in S} u(x_S^i) = n \sum_{i\in S} \frac{1}{n} u(x_S^i) \leq n \cdot u\left(\sum_{i\in S} \frac{1}{n} x_S^i\right) = n \cdot \frac{1}{n} u\left(\sum_{i\in S} x_S^i\right) \quad.$$

Hence

$$\sum_{i \in S} u(x_S^i) \leq u(\sum_{i \in S} x_S^i) \quad ,$$

or

$$w(S) \leq u(\chi_S) \quad ,$$

by the definitions of $w(S)$ and $u(\chi_S)$. Let us now prove that $u(\chi_S) \leq v(S)$. We have

$$u(\chi_S) = \max_{\{\alpha_T\}} \{ \sum_{T \subseteq N} \alpha_T v(T) : \alpha_T \geq 0 \quad , \quad \sum_{T \subseteq N} \alpha_T \chi_T = \chi_S \} \quad .$$

Let the maximum be $\sum_{T \subseteq N} \hat{\alpha}_T v(T)$, and consider the subgame corresponding to S. Since $\sum \hat{\alpha}_T \chi_T = \chi_S$, all the members of T of every feasible collection are subsets of S. Therefore if we consider each T as a coalition for the subgame,

$$\sum \hat{\alpha}_T \chi_T = 1 \quad .$$

Thus the collection of T's is balanced in the subgame, with balancing weights $\{\hat{\alpha}_T\}$. So by the Bondareva-Shapley theorem applied to the subgame

$$\sum \hat{\alpha}_T v(T) \leq v(S) \quad ,$$

or

$$u(\chi_S) \leq v(S) \quad .$$

Above it was established that $w(S) \leqq u(\chi_S)$, so we have $w(S) \leqq v(S)$. This, together with the conclusion above that $w(S) \geqq v(S)$ yields

$$v(S) = w(S) \quad \text{for all} \quad S \subset N \quad .$$

The initial game (N,v) is then a market game (since (N,w) is) and this completes the proof of the theorem.

Chapter 6: The von Neumann-Morgenstern Solution

The "von Neumann-Morgenstern solution" was the first solution concept to be studied (see von Neumann [1928]). It was later extensively examined by von Neumann and Morgenstern [1944], and by subsequent workers. The ideas on which it is founded are closely related to those on which the core is founded, and it will be introduced here on the basis of these ideas. Throughout we will use $x(S)$ to denote $\sum_{i \in S} x^i$.

6.1 <u>Definition</u>: Let x and y be payoff vectors, and let S be a coalition. x <u>dominates</u> y <u>via</u> S (written $x \underset{S}{\succ} y$) if

$$x^i > y^i \quad \text{for all} \quad i \quad \text{in} \quad S$$

and

$$x(S) \leqq v(S) \quad .$$

x <u>dominates</u> y (written x $\succ$ y) if there is an S such that x $\underset{S}{\succ}$ y.

6.2 <u>Lemma</u>: <u>An imputation</u> y <u>is in the core if and only if it is</u> <u>not dominated by any payoff vector</u>.

<u>Proof</u>: Let y be in the core. If x $\underset{S}{\succ}$ y, then $v(S) \geq x(S) > y(S)$. But $y(S) \geq v(S)$ since y is in the core. Thus we have a contradiction, and there is no x which dominates y.

Conversely, suppose y is an imputation not in the core. Then there is an S such that $y(S) < v(S)$. Define a payoff vector x by

$$x^i = \begin{cases} y^i + \dfrac{v(S) - y(S)}{|S|} & , \text{ if } i \in S \\ 0 & \text{otherwise .} \end{cases}$$

Then $x^i > y^i$ for all $i \in S$ and

$$x(S) = y(S) + |S| \frac{v(S) - y(S)}{|S|} = v(S) \quad .$$

So x $\underset{S}{\succ}$ y. This proves the Lemma.

6.3 <u>Lemma</u>: <u>Assume that</u> v <u>is superadditive</u>. <u>Then an imputation</u> y <u>is in the core if and only if it is not dominated by any imputation</u>.

<u>Proof</u>: The necessity of the condition follows immediately from Lemma 6.2 above.

For the sufficiency of the condition, let y and S be as in the sufficiency proof of Lemma 6.2. Define a payoff vector x by

$$
x^i = \begin{cases} y^i + \dfrac{v(S) - y(S)}{|S|} & \text{if } i \in S \\[3mm] v(i) + \dfrac{v(N) - [v(S) + \sum\limits_{i \notin S} v(i)]}{|N \setminus S|} & \text{if } i \notin S . \end{cases}
$$

Because of superadditivity, $v(N) - [v(S) + \sum\limits_{i \in S} v(i)] > 0$, so x is individually rational; also $\sum\limits_{i \in N} x^i = v(N)$, so x is an imputation. Finally, $x(S) = v(S)$ and $x \underset{S}{\succ} y$, so the Lemma is proved.

Exercise 8: Show that without the assumption of superadditivity Lemma 6.3 is false.

The stability concept underlying the definition of the core could be criticized as being too strong. It does not seem natural to exclude as unstable a dominated payoff vector when the dominating payoff vector is itself not stable. This suggests that we should focus our attention on domination by stable imputations.

6.4   Definition: A set K of imputations is called a von Neumann-Morgenstern (N-M) solution (or simply a solution) of v if K is the set of all imputations not dominated by any member of K.

6.5   Remark: K in the definition above may not be unique and may not exist (even as the empty set).

6.6   Proposition: K is a solution of v if and only if for all imputations x and y:

1. If $x, y \in K$, then x does not dominate y. (Internal consistency.)

2. If $y \notin K$, there is an $x \in K$ that dominates y. (External domination.)

The following notation allows the proposition to be stated more compactly:

Dom x = the set of all payoff vectors dominated by x;

Dom K = the set of all imputations dominated by a member of K

$= \cup_{x \in K}$ Dom x;

X = the set of all imputations.

Conditions 1 and 2 then become

1. $K \subseteq (X \backslash Dom\ K)$

2. $K \supseteq (X \backslash Dom\ K)$.

What the proposition says, then, is that

K is a solution if and only if $K = X \backslash Dom\ K$.

The proof of the proposition is immediate, using Definition 6.4.

6.7   Remark: The core is a subset of every N-M solution.

Let us determine the N-M solution for some simple games.

6.8   Example: 2-person bargaining game. We have $N = \{1,2\}$, $v(N) = 1$, and $v(1) = v(2) = 0$. We recall that the core (see 4.7) is the set of all imputations

$$X = \{(x^1, x^2): \quad x^1 \geq 0, \; x^2 \geq 0, \; x^1 + x^2 = 1\} \quad .$$

Since the core is a subset of every N-M solution, if a solution exists it can only be X. To prove that X is indeed a solution, we prove that Dom X = ∅, so that X = X\Dom X.

Assertion: In any game, no imputation dominates another one via a one-person coalition.

Proof: Consider a 0-1 normalized game (we can do so w.l.o.g.). Suppose $x \underset{\{i\}}{\succ} y$. Since $0 \leq x^i \leq v(i) = 0$ and $y^i \geq 0$, we must have both $x^i = 0$ and $x^i > y^i$, which is a contradiction.

So a one-person coalition can never dominate. Hence in the 2-person bargaining game the only possible domination is via the coalition {12}.

Assertion: In any game, no imputation dominates another one via the coalition N of all players.

Proof: Let x and y be two imputations. Then $x(N) = v(N)$ and $y(N) = v(N)$. But $x \underset{N}{\succ} y \Rightarrow x(N) > y(N)$, so we have a contradiction.

Hence in the two-person bargaining game no imputation is dominated, so that there is one and only one solution, namely the set of all imputations.

6.9 Example: 3-person majority game. In this game $N = \{1,2,3\}$, $v(N) = v(12) = v(13) = v(23) = 1$, and $v(i) = 0$ for all $i \in N$.

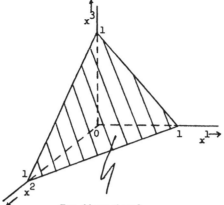

Two-dimensional
simplex

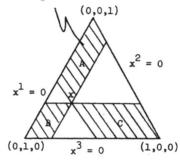

We recall that the set of imputations is the two-dimensional simplex, shown in the diagrams above. First consider which imputations are dominated by x. By the two assertions above, domination can only be via the coalitions {12}, {23}, and {13}. Assume y is dominated by x via {12}. Then

$$x^1 > y^1 \quad , \quad x^2 > y^2 \quad , \quad \text{and} \quad x(12) \leq v(12) = 1 \quad .$$

Hence any point in the area A of the diagram is dominated in this
way (the "outside" sides of A are included, but the "inside" sides
are excluded). Using a symmetrical argument points in areas B and
C are dominated by x via the coalitions {13} and {23}. From
this it can be seen that no single point can dominate all others, so that
no singleton can constitute a N-M solution.

It can also be seen that if K is a solution, then if x and
y are in K, the line joining x and y must be parallel to one of
the sides of the triangle of imputations (otherwise one point dominates
the other). It appears, then, that either a line, or the vertices of
an appropriate triangle qualify as possible solutions.

Consider first the vertices of a triangle. Let them be a, b
and c in the diagram below. The shaded area consists of points

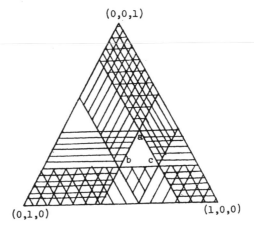

dominated by a, b, or c. From this it is clear that no such three points can constitute a solution: there will always be points not dominated by any of them.

Consider now the situation if the vertices a, b, and c of the triangle are oriented as in the diagram below. Points in the shaded area are dominated. So if a, b, and c are as in the next diagram, they will dominate all of $X \setminus \{a,b,c\}$. Hence a solution is

$$K = \{(0,\tfrac{1}{2},\tfrac{1}{2}),(\tfrac{1}{2},0,\tfrac{1}{2}),(\tfrac{1}{2},\tfrac{1}{2},0)\} \quad .$$

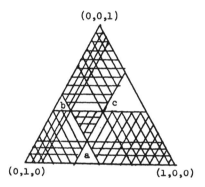

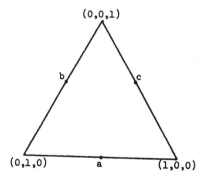

Now consider the possibility that a line segment parallel to one of the sides of the triangle is a solution. An example is the line joining d and e in the diagram below. It can be seen that

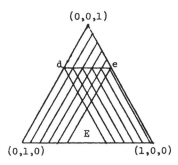

for points on the line to dominate all other imputations the set E must be empty: the line must be at least half way down the triangle. Hence any set

$$K_1 = \{(x, 1 - x - c, c) : 0 \leq x \leq 1 - c\} \quad \text{for} \quad 0 \leq c < \frac{1}{2}$$

is an N-M solution. By symmetry, the same is true of the sets

$$K_2 = \{(c, x, 1 - x - c) : 0 \leq x \leq 1 - c\}$$

and

$$K_3 = \{(1 - x - c, c, x) : 0 \leq x \leq 1 - c\}$$

for $0 \leq c < 1/2$.

To sum up, there are two different types of solutions.

(1) The symmetric solution $K = \{(0,1/2,1/2),(1/2,0,1/2),(1/2,1/2,0)\}$ Here two out of the three players get together in a coalition and divide the payoff equally between themselves; the three players are symmetric.

(2) Discriminatory solutions. Two players get together, give an amount less than half to the third player, and bargain over the remaining payoff. Here one individual is ostracized; i.e. excluded from the bargaining process.

Two features of these solutions are noteworthy:

(1) An N-M solution can be interpreted as a stable form of organization for society. Here, two forms of organization in which the same people are treated differently are both stable.

(2) The behavior of the people involved in bargaining is qualitatively different in the two forms of organization. When the three people are in symmetric positions and two get together, neither one will settle for less than 1/2 since each can say: "If I don't get my due share, I will go along with the third player and get it from him." In the discriminatory case, the bargaining process is different: by common consent the third player is ostracized.

Let us now examine the N-M solution for a more general class of majority games.

6.10 <u>Definition</u>: A weighted majority game with weights $\{w^i\}_{i \in N}$ and quota q is called <u>strong</u> if for all S, either S or N\S is winning, but not both.

6.11 Definition: A minimal winning coalition is a coalition S such that no strict subset of S is winning.

6.12 Remark: Note that a given weighted majority game (N,v) does not determine unique weights and quota. For instance, $[q; w^1, w^2, w^3]$ = [5; 2,3,4] generates the same game as [2; 1,1,1] (namely the 3-person majority game).

6.13 Definition: A representation of a weighted majority game (N,v) is a set of non-negative numbers $[q; w^1, w^2, \ldots, w^n]$ such that $v(S) = 1$ if and only if $\sum_{i \in S} w^i \geq q$.

6.14 Definition: A weighted majority game is called homogeneous if it has a representation in which $\sum_{i \in S} w^i = q$ for all minimal winning coalitions S.

6.15 Remark: A strong weighted majority game may not be homogeneous.

Exercise 9: Consider the weighted majority game determined by the representation [5; 2,2,2,1,1,1]. {1,2,3} is a minimal winning coalition, but its total weight 6 exceeds the quota 5. Prove that there is no representation which makes this game homogeneous.

6.16 Theorem (von Neumann and Morgenstern [1944]): Consider a strong homogeneous weighted majority game with a homogeneous representation $[q; w^1, \ldots, w^n]$. Let q = 1. For each minimal winning coalition S, define

$$y_S^i = \begin{cases} w^i & \underline{if} \quad i \in S \\ 0 & \underline{otherwise} \end{cases} .$$

<u>Let</u> $K = \{y_S: \ S \ \text{is a minimal winning coalition}\}$. <u>Then</u> K <u>is a</u> <u>solution</u>.

Proof: We first prove the internal consistency of K. Let S and T be minimal winning coalitions. Assume that

$$y_S \underset{U}{\succ} y_T \ .$$

U will have to be a subset of S since $y_S^i = 0$ for $i \notin S$. If U is a proper subset of S then it is a losing coalition, so $v(U) = 0$ and hence

$$y_S^i = 0 \quad \text{for all} \ i \ \text{in} \ U,$$

in which case y_S could not dominate y_T via U. Hence $U = S$.

Suppose now that $T \cap S \neq \emptyset$. Let $j \in T \cap S$. Then

$$y_S^j = w^j = y_T^j \ ,$$

in which case once again y_S could not dominate y_T. Hence $T \cap S = \emptyset$. But since the game is strong, two disjoint winning coalitions cannot exist. Hence there is no U such that $y_S \underset{U}{\succ} y_T$, and the set K is internally consistent.

We now prove that the external domination condition is satisfied. Let z be any imputation. Let T be the set of all i such that $w^i > z^i$. If T is winning, then T has a minimal winning subset S and $y_S \underset{S}{\succ} z$, with $y_S \in K$. If T is losing, then $\mathbb{N}T$ is winning ($\mathbb{N}T = \{i : w^i \leq z^i\}$). Let S be a minimal winning subset of $\mathbb{N}T$. Then

$$1 = q = \sum_{i \in S} w^i \leq \sum_{i \in \mathbb{N}T} w^i \leq \sum_{i \in \mathbb{N}T} z^i \leq \sum_{i \in \mathbb{N}T} z^i + \sum_{i \in T} z^i = \sum_{i \in N} z^i = 1$$

(using the fact that the game is homogeneous). So $\sum_{i \in T} z^i = 0$ and

$$z^i = \begin{cases} w^i & \text{for } i \in S \\ 0 & \text{for } i \notin S \end{cases}.$$

We conclude that $z = y_S$. Hence either $z \in K$ or there exists $y_S \in K$ such that $y_S \underset{S}{\succ} z$. Hence the condition of external domination is satisfied.

This completes the proof of the theorem.

The solution in this case can be interpreted in the following way: a minimal winning coalition forms and its members divide the payoff according to the homogeneous weights.

6.17 **Example**: Market with one buyer and two sellers. In this game

$$N = \{1,2,3\} \quad , \quad v(N) = v(12) = v(13) = 1 \quad ,$$

and

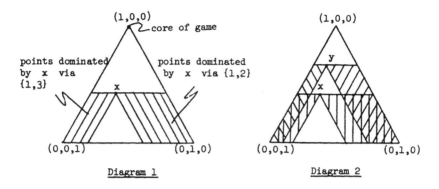

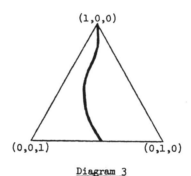

Diagram 3

$$v(23) = v(1) = v(2) = v(3) = 0 \quad .$$

Since $v(23) = 0$, and because of the assertions made in Example 6.8, domination can only be via the coalitions $\{1,2\}$ and $\{1,3\}$. Also, we know
that the core of this game is $\{(1,0,0)\}$ (see Example 4.9), and that this
is contained in every solution. Once again the set of imputations is
represented by the triangle shown in Diagram 1 (it has been re-oriented

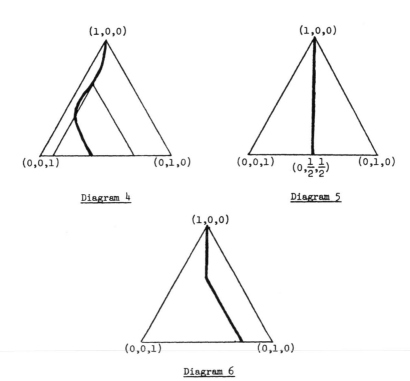

Diagram 4

Diagram 5

Diagram 6

for elegance in presentation). The set of imputations dominated by the imputation x is shaded. In Diagram 2 the set of imputations dominated by the imputations x and y is shaded. From this it can be seen that in order to satisfy the condition of external domination one needs every point on some curve from the point (1,0,0) to the bottom of the triangle. At the same time, in order to satisfy the condition of internal consistency, it must be the case that all points on the curve below any given point lie between the two straight lines through z parallel to the sloping sides of the triangle. Thus the curve in Diagram 3 is a solution, while the curve in Diagram 4 is not.

It is possible to give an interpretation to these solutions. First consider the case given in Diagram 5. Here one can argue that the two sellers form a cartel, bargaining as a single unit with the buyers, and splitting the payoff they extract from him equally. Cases in which the solution is curvilinear can be interpreted as a situation where the sellers form a cartel, but split their payoff according to some nonlinear scheme. For example, the solution in Diagram 6 represents a situation where players 2 and 3 split the payoff to their cartel equally if it is less than some number, with all of any excess above this quantity going to player 2. The restriction on the shape of a curve which is a solution means, in this interpretation, that the payoffs to players 2 and 3 must each be nondecreasing in the payoff to the cartel.

In this example, then, one can interpret the solutions as predicting the formation of a cartel. We will now consider a whole class of games for which some sort of cartelization is predicted. First, some definitions which will be used are presented.

6.18   Definition: A permutation of the players is a one-one mapping π from N to N.

6.19   Definition: A set K of payoff vectors is symmetric if for each $x \in K$ and each permutation π of the players, $\pi x \in K$ where $(\pi x)^i \equiv x^{\pi(i)}$ for all $i \in N$.

6.20   Definition: An imputation $x \in E^n$ is monotonic if $x^i \geqq x^{i+1}$
for all $i = 1,\ldots,n - 1$.

6.21   Example: (n,k) games. These games are defined by $|N| = n$ and

$$v(S) = \begin{cases} 1 & \text{if } |S| \geqq k \\ 0 & \text{otherwise ,} \end{cases}$$

with $k \geqq (n + 1)/2$. (It can be seen that an (n,k) game is a non-
strong weighted majority game.) We will, for simplicity, concentrate
on the game for which $n = 10$ and $k = 8$; our considerations generalize
without difficulty to the general case (see Bott [1953]).

Assertion: A solution of the $(10,8)$ game is given by

$$K = \{x: \ \pi x = (a^1,a^1,a^1,a^2,a^2,a^2,a^3,a^3,a^3,0) \quad \text{for some } \pi,$$

$$a^i \in [0,1] \quad \text{for } i = 1,2,3, \text{ and } 3 \sum_{i=1}^{3} a^i = 1 \} \ .$$

Proof: Since K is symmetric, it contains a monotonic imputation and we can confine our attention to such imputations. Let x
and y be monotonic imputations. Then it is clear that in this case
x dominates y if and only if there exists a minimal winning coalition
S such that $x^i > y^i$ for all $i \in S$. Hence there is some permutation
of x which dominates y if and only if the first eight members of x
are larger respectively than the last eight members of y.

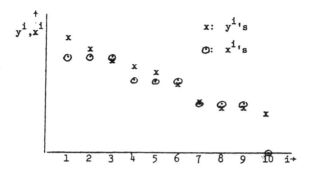

(a) <u>External Domination</u>: Let y be a monotonic imputation not in K; such a y is shown in the diagram above. By the remark above, we have to find an x in K such that

$$x^i > y^{i+2} \quad \text{for } i = 1,\ldots,8 ,$$

in order to show that y is dominated by an imputation in K. Let

$$e = \sum_{i=1}^{2} (y^i - y^3) + \sum_{i=4}^{5} (y^i - y^6) + \sum_{i=7}^{8} (y^i - y^9) + y^{10} .$$

Since $y \notin K$, e > 0. Let

$$x = (y^3 + \tfrac{e}{9}, y^3 + \tfrac{e}{9}, y^3 + \tfrac{e}{9}, y^6 + \tfrac{e}{9}, y^6 + \tfrac{e}{9}, y^6 + \tfrac{e}{9}, y^9 + \tfrac{e}{9}, y^9 + \tfrac{e}{9}, y^9 + \tfrac{e}{9}, 0)$$

Then $\Sigma x^i = 3y^3 + 3y^6 + 3y^9 + e = \Sigma y^i = 1$, so x is an imputation; it is clearly contained in K, and $x^i > y^{i+2}$ for $i = 1,\ldots,8$. Hence the condition of external domination is satisfied.

(b) <u>Internal Consistency</u>: Suppose x and y are monotonic imputations in K and some permutation of x dominates y. Then $x^i > y^{i+2}$ for $i = 1,\ldots,8$. But then $x^1 > y^3$, $x^4 > y^6$ and $x^7 > y^9$, so that $x^1 > y^1$, $x^4 > y^4$, and $x^7 > y^7$, in which case $\Sigma x^i > \Sigma y^i$. Hence x cannot be an imputation, and this contradiction establishes internal consistency.

The proof of the assertion is now complete.

It can in fact be shown that K is the unique symmetric solution of the $(10,8)$ game (see Bott [1953]).

6.22 <u>Definition</u>: In a simple game S is <u>blocking</u> if $N \backslash S$ is losing.

In the above example an interpretation is that the players get together in minimal blocking conditions. In strong weighted majority games it was found above (Theorem 6.16) that the solution predicted the formation of minimal <u>winning</u> coalitions. However, in such games a coalition is minimal winning if and only if it is blocking, so that Example 6.21 indicates that the significant aspect of these coalitions is, in fact, that they are minimal blocking.

Chapter 7: Bargaining Sets

7.1 <u>Example</u>: Consider the game (N,v) where

$N = \{1,2,3\}$, $v(i) = 0$ for all i ,

$v(12) = v(13) = v(123) = 1$, and $v(23) = \frac{1}{2}$.

While 2 and 3 are in symmetric positions in this game, it appears
that 1 is in a stronger position. Two problems can be considered: what
coalitions will form?; and how will the members of the coalitions so formed
divide their worth among themselves? There is no uniquely "correct" way
of dealing with these problems, but the "Bargaining Set" represents one
approach to the second problem, taking the coalitions which form as given.
Consider the case where 1 and 2 get together in a coalition. Suppose
that they are considering the payoff vector $(2/3,1/3,0)$. Player 1 can
say that this is not satisfactory since he could get together with player 3
and establish the payoff vector $(5/6,0,1/6)$, which would benefit both
himself and player 3. But player 2 can reply that he could also offer 1/6
to player 3, establishing the payoff vector $(0,1/3,1/6)$, where he is as well
off as he was before, and 3 is as well off as he would be in player 1's
proposed deviating payoff vector. However, player 1 could propose establishing
$(0.8,0,0.2)$ together with player 3, so that if player 2 were to give 3
as much as he gets in this payoff vector he would have to get less himself
than he did in the original payoff vector which was being considered:
$v(23) = 1/2$, so the most 2 could get if he gave 0.2 to player 3,
would be 0.3, while originally he got 1/3. In this way the superior
"strength" of player 1 is revealed, and he might suggest that $(0.7,0.3,0)$
represents a reasonable split of the proceeds between himself and player 2.
But exactly as above, 1 could threaten with $(0.74,0,0.26)$, for example,
which 2 could not match. So it appears that player 1 will receive an
even larger payoff. Consider, then, the payoff vector $(0.8,0.2,0)$ as
a candidate for agreement between 1 and 2. In this case it is possible

for player $\underline{2}$ to threaten with $(0,1/4,1/4)$, a threat which player $\underline{1}$ is unable to match, since if he gives 3 at least 1/4, at most 3/4 (< 0.8) will be left for himself. Consider, however, the payoff vector $(3/4,1/4,0)$; if player 1 threatens to join with 3, at the same time increasing his payoff from 3/4, he will have to give 3 at most 1/4, while player 2 can always counter such a move by threatening to join with 3, giving him 1/4 while maintaining his own payoff. Similarly, if player 2 threatens to join with 3, and at the same time increases his payoff, then he can give 3 at most 1/4, while player 1 can always counter such a move by threatening to join with 3, giving him 1/4, and maintaining his own payoff at 3/4. In this way neither 1 nor 2 can object in a convincing way to the payoff vector $(3/4,1/4,0)$, and the arguments above indicate that this is the only payoff vector for which this is so (for the grouping of players under consideration): it is, in fact, the unique member of the "Bargaining Set" in the case where players 1 and 2 get together in a coalition.

The arguments used above are similar to those used when the question of whether a payoff vector is in the core is being considered; but there, only the original threats to deviate are considered; the reasoning behind the "Bargaining Set" goes beneath the surface of this sort of argument and considers the possibility that threats by some players are "counterbalanced" by threats from other players. Thus, the core is the set of payoff vectors to which there is no objection, while the "Bargaining Set" is the set of payoff vectors to which there is no justified objection. We can now define these notions precisely.

7.2 **Definition:** A <u>coalition structure</u> is a partition $B = \{B_1, B_2, \ldots, B_k\}$ of the set N of players (i.e. $B_i \cap B_j = \emptyset$ for $i \neq j$ and $\bigcup_{i=1}^{k} B_i = N$).

7.3 **Definition:** $X_B = \{x \in E^N : x(B_i) = v(B_i)$ for $i = 1, \ldots, k$ and $x^i \geq v(\{i\})$ for all $i \in N\}$.

7.4 **Remark:** The condition $x(B_i) = v(B_i)$ for $i = 1, \ldots, k$ means that the total worth of each coalition B_i is completely divided up between the members of the coalition, and $x^i \geq v(\{i\})$ is the individual ratio-nality condition. In the example above, $B = \{12, 3\}$; another example is $B = \{N\}$, in which case $X_B = X_{\{N\}} = \{x \in E^N : x(N) = v(N)$ and $x^i \geq v(\{i\})$ for $i \in N\}$. Thus $X_{\{N\}}$ coincides with the set of imputations.

7.5 **Definition:** Given a game (N,v), a coalition structure B, a payoff vector $x \in X_B$, a set $B_k \in B$ and two members i and j of B_k, an <u>objection of i against j</u> consists of a set S containing i but not j, and a point $y \in E^S$ such that $y^i > x^i$, $y^\ell \geq x^\ell$ for all $\ell \in S$, and $y(S) \leq v(S)$.

7.6 **Remark:** The interpretation is that i gets together with a group of players not including j and realizes a payoff vector in which he obtains more than he is getting at present, while the other members of the group get at least as much as they are at present getting.

7.7 **Definition:** A <u>counterobjection to y by j</u> consists of a set T containing j but not i, a point $z \in E^T$ such that $z^j \geq x^j$, $z^\ell \geq x^\ell$ for all $\ell \in T$, $z^\ell \geq y^\ell$ for all $\ell \in S \cap T$ and $z(T) \leq v(T)$.

An objection is _justified_ if there is no counterobjection to it. The payoff vector x is in the _Bargaining Set_ $M = M(v, \mathcal{B})$ if there is no justified objection to it.

Thus, in the example above, $M(v, \{12,3\}) = \{(3/4, 1/4, 0)\}$. An important question is whether the Bargaining Set is nonempty for any $\mathcal{B}$ when $X_\mathcal{B} \neq \emptyset$. Peleg [1963] and [1967] solved the problem originally, using a fixed point argument. Maschler and Peleg [1966] later found a completely different algebraic existence proof, and later still Schmeidler [1969] devised a still simpler proof, which is followed here. Before that is presented, consider another example.

7.8   _Example:_ Consider the weighted majority game $[3; 2,1,1,1]$ (i.e. $v(S) = \{{}^{1}_{0} \; {}^{\text{if } w(S) \geq 3}_{\text{otherwise}}$ where $(w^1, w^2, w^3, w^4) = (2,1,1,1))$. Suppose the coalition structure is $\mathcal{B} = \{12, 3, 4\}$. It is clear that every point in the Bargaining Set will be of the form $(\alpha, 1 - \alpha, 0, 0)$. Objections of 1 are of the form $(\alpha + \varepsilon, 0, 1 - \alpha - \varepsilon, 0)$, and the smaller ε, the better they are (i.e. the more difficult to counterobject to). For player 2 to be able to counterobject he must join with 3 and 4, give himself $1 - \alpha$, and have at least $1 - \alpha - \varepsilon$ left over to give to player 3 (he need give player 4 nothing, since that is what he is getting at present). I.e. it must be the case that

$$v(234) - (1 - \alpha) \geq 1 - \alpha - \varepsilon \qquad \text{for all } \varepsilon > 0 \; ,$$

or

$$1 - 1 + \alpha \geq 1 - \alpha - \varepsilon \quad \text{for all } \varepsilon > 0 \; , \quad \text{or } \alpha \geq \frac{1}{2} \; .$$

Now consider an objection by 2. If he objects with $(0,1-\alpha-\epsilon,\alpha-\epsilon,$ player 1 can easily counterobject by getting together with 4; a "good" objection by 2 is $(0,1-\alpha-\epsilon,(\alpha-\epsilon)/2,(\alpha-\epsilon)/2)$. Then player 1 can counterobject if $v(13) - \alpha \geq (\alpha-\epsilon)/2$ for all $\epsilon > 0$ (he only has to get together with one player to obtain a worth of 1), or if $1-\alpha \geq (\alpha-\epsilon)/2$ for all $\epsilon > 0$, or, if $\alpha \leq 2/3$. This exhausts the possibilities for objection so there are no justified objections to $(\alpha,1-\alpha,0,0)$ if $\alpha \in [1/2,2/3]$. Hence the Bargaining Set is $M(v,\{12,3,4\}) = \{(\alpha,1-\alpha,0,0)\}_{1/2 \leq \alpha \leq 2/3}$.

Exercise 10: Find $M(v, \{1234\})$ for the weighted majority game [3; 2,1,1,1].

Exercise 11: Find $M(v, \{123\})$ for the game defined in Example 7.1.

Now, instead of considering the details of the procedure involved in establishing whether or not a payoff vector is in the Bargaining Set, one could deal with a "rough" measure of the "strength" of a player (his ability to object to a payoff vector x, or to counterobject to an objection to x), as follows:

7.9 Definition: $v(S) - x(S)$ is the excess of the coalition S. Given v, B, $B_k \in B$, i and j in B_k, $S \ni i$, and $j \notin S$, $s_{ij}(x) \equiv \max \{v(S) - x(S): S \ni i \text{ and } S \not\ni j\}$ is the maximum excess of i against j.

7.10 Definition: The Kernel, K, (given v and B) is the set of all payoff vectors $x \in X_B$ such that for all $B_k \in B$, and i and j in B_k, either $s_{ij} \leq s_{ji}$ or $x^j = v(\{j\})$.

7.11   Remark: x will be in the Kernel if for all i and j either
$s_{ij} \leqq s_{ji}$ -- i.e., i's strength vis-a-vis j is not greater than j's
strength vis-a-vis i -- or it is greater, but j is at his personal
minimum in any case, so that i cannot convincingly suggest that j's
payoff be reduced.

7.12   Theorem (Davis and Maschler [1965]): The Kernel is a subset of the
Bargaining Set.

   Proof: Let $x \in$ Kernel. Let $i, j \in B_k \in B$ and let y, S be
an objection of i against j. Then $y(S) \leqq v(S)$, and from the definition
of s_{ij}, $v(S) - x(S) \leqq s_{ij}$. So:

   (a) If $x^j = v(\{j\})$, then j can counterobject by himself--
i.e. $(0,\ldots,0,v(\{j\}),0,\ldots,0),\{j\}$ is a counterobjection to y by j.

   (b) Otherwise $x^j > v(\{j\})$ and $s_{ji} \geqq s_{ij} \geqq v(S) - x(S)$; but
$s_{ji} \equiv \max \{v(S) - x(S): S \ni j, S \not\ni i\}$, so there exists T with $j \in T$,
$i \notin T$, such that $v(T) - x(T) \geqq v(S) - x(S)$. Also, $v(S) - x(S) \geqq y(S) - x(S)$
from the above. So there exists T such that $v(T) - x(T) \geqq y(S) - x(S)$,
or $v(T) \geqq y(S) + x(T) - x(S) = y(S) - x(S) + x(T\backslash S) + x(T \cap S)$.
But $y(S) - x(S) \geqq y(S \cap T) - x(S \cap T)$ since $y(S) > x(S)$. So
$v(T) \geqq y(S \cap T) + x(T\backslash S)$. Hence j can give to everyone in $T \cap S$
at least as much as they get in y, and to those in $T\backslash S$ at least as
much as they get in x. Hence j can counterobject.

   Hence in all cases j can counterobject to any objection of i --
so any point in the Kernel is certainly in the Bargaining Set. So the
theorem is proved.

7.13 <u>Example</u>: Let us compute the Kernel for the coalition structure
$B = \{12,3,4\}$ in the case of the weighted majority game considered above
$([3; 2,1,1,1])$. We know that the Bargaining Set $M(v,\{12,3,4\}) =$
$\{(\alpha, 1 - \alpha, 0, 0)\}_{1/2 \leq \alpha \leq 2/3}$. The Kernel K is a subset of M. When does
an imputation $x = (\alpha, 1 - \alpha, 0, 0)$ belong to the Kernel? We have to
compute $s_{12}(x)$ and $s_{21}(x)$. We know $v(1) - x(1) = -\alpha$, $v(13) - x(13) =$
$1 - \alpha$, $v(14) - x(14) = 1 - \alpha$, and $v(134) - x(134) = 1 - \alpha$. Hence
$s_{12}(x) \equiv \max \{v(S) - x(S): S \ni i, S \not\ni j\} = 1 - \alpha$. In the same way,
$s_{21}(x) = \alpha$. Hence one possible point in the Kernel is an x such that
$s_{12}(x) = s_{21}(x)$, which leads to $\alpha = 1/2$. It is the only point since the
condition $x^j = v(\{j\})$ leads to two imputations $(0,1,0,0)$ and $(1,0,0,0)$
outside the Bargaining Set, and <u>a fortiori</u> outside the Kernel. So
$K = \{(1/2, 1/2, 0, 0)\}$. Note that this point is in no sense the "center" of
the Bargaining Set, and that, in particular, the advantage of player 1
over player 2, which is reflected in the Bargaining Set, does not show
up in the Kernel.

The concept of excess was implicit in the definition of the
Bargaining Set, and explicit in the definition of the Kernel. In both
definitions, however, there is the idea of an underlying specific bargaining process between agents i and j. We are going to introduce a
new solution concept, the Nucleolus, which abandons the idea of dialogue
between i and j and for which the concept of excess, as a measure
of objecting power, is central. Let us consider a game (N,v) with a
coalition structure B. For a given $x \in X_B$ there exist 2^n excesses
$\{v(S) - x(S)\}_{S \subset N}$. Given x, let us index the coalitions $S \subset N$ so that

$$v(S_1) - x(S_1) \geqq v(S_2) - x(S_2) \geqq \ldots \geqq v(S_{2^n}) - x(S_{2^n}) \quad .$$

and let us define $\theta(x) = (v(S_1) - x(S_1), v(S_2) - x(S_2), \ldots, v(S_{2^n}) - x(S_{2^n})) \in E^{2^n}$.
Let θ_i be the i-th coordinate of θ.

7.14 $\underline{Definition}$: Let E^r be a Euclidean space, and x and y two points in E^r. We define the $\underline{lexicographic\ order}$ $\succcurlyeq$ on E^r in the following way: $x \succcurlyeq y$ if there exists i such that $x_j = y_j$ for all $j < i$ and $x_i > y_i$. x is then said to be $\underline{lexicographically\ greater}$ than y.

7.15 $\underline{Definition}$: For a game (N,v) and for a given coalition structure B, the $\underline{Nucleolus}$ $Nu(N,v,B)$ is the set of all x in X_B such that there is no y in X_B with $\theta(x) \succcurlyeq \theta(y)$. Hence a point in the Nucleolus is a lexicographic minimum of θ over X_B.

7.16 $\underline{Remark}$: The Nucleolus may be interpreted in the following way: the excess measures the "dissatisfaction" of a coalition S with the proposed accommodation x; $v(S) - x(S)$ is the difference between what the coalition could get alone and what it would get if the accommodation were actually implemented. B represents a given structure of society; any payoff vector x in X_B fully "satisfies" any coalition in B. If we think of the loudness of S's complaint against x as proportional to its dissatisfaction, the Nucleolus may be considered as the result of the following process: the "judge" (or the "government") minimizes the loudest complaint; subject to achieving this, he minimizes the

second loudest complaint; and so forth. Then the Nucleolus is the set
of points at which the overall loudness of complaints is a minimum (in
the lexicographic sense), given the structure of society. The idea of
individual bargaining, essential in the concepts of the Kernel and the
Bargaining Set, is not present here.

Now, we want to prove the nonemptiness of the Bargaining Set. We
will prove that the Nucleolus is nonempty and that the Nucleolus is a
subset of the Bargaining Set.

7.17 Theorem (Schmeidler [1969]): The Nucleolus $Nu(N,v,B)$ is nonempty
(if X_B is nonempty).

Proof: First we establish the following.

Lemma (Schmeidler [1969]): Let $f_1, f_2, \ldots, f_r$ be r continuous
functions on some space. For a given x, define $i_k(x)$ for $k = 1, \ldots, r$
such that $f_{i_1(x)}(x) \geq f_{i_2(x)}(x) \geq \cdots \geq f_{i_k(x)}(x) \geq \cdots \geq f_{i_r(x)}(x)$.
Then $f_{i_k(x)}(x)$ is a continuous function of x for all $k = 1, \ldots, r$.

Proof: $f_{i_1(x)}(x) = \max [f_1(x), f_2(x), \ldots, f_r(x)]$. Hence $f_{i_1(x)}(x)$
is continuous, being the maximum of a finite number of continuous functions.

$$f_{i_2(x)}(x) = \min \{\max [f_2(x), f_3(x), \ldots, f_r(x)], \max [f_1(x), f_3(x), \ldots, f_r(x)], \ldots,$$
$$\max [f_1(x), f_2(x), \ldots, f_{r-1}(x)]\} \quad ,$$

so $f_{i_2(x)}(x)$ is the minimum of r functions which are continuous, each
being the maximum of $r - 1$ continuous function; hence it is continuous.

Similarly, $f_{i_k(x)}(x)$ is continuous for $k = 1,\ldots,n$. This proves the lemma.

Now we have the following.

(a) $\theta_1(x)$ is continuous and X_B is a compact set (since it is defined by a finite number of weak inequalities). Let

$$\zeta_1 = \min \{\theta_1(x) : x \in X_B\} \quad \text{and} \quad X_1 = \{x \in X_B : \theta_1(x) = \zeta_1\} \ .$$

Then X_1 is nonempty.

(b) $\theta_2(x)$ is continuous by the Lemma above. X_1 is closed, being the inverse image of a closed set $\{\zeta\}$ under a continuous function; it is compact, being a closed subset of a compact set. So, letting

$$\zeta_2 = \min \{\theta_2(x) : x \in X_B\} \quad \text{and} \quad X_2 = \{x \in X_1 : \theta_2(x) = \zeta_2\} \ ,$$

X_2 is seen to be nonempty.

(c) Similarly for $i = 3,\ldots,2^n$, $\theta_i(x)$ is continuous. Let

$$\zeta_i = \min \{\theta_i(x) : x \in X_B\} \quad \text{and} \quad X_i = \{x \in X_{i-1} : \theta_i(x) = \zeta_i\} \ .$$

Then, as above, if $X_{i-1} \neq \emptyset$, $X_i \neq \emptyset$. So by induction, $X_i \neq \emptyset$ for all $i = 1,\ldots,2^n$. So $Nu = X_{2^n}$ is nonempty, and the theorem is true.

7.18 <u>Theorem</u> (Schmeidler [1969]): <u>The Nucleolus</u> Nu <u>is a subset of the Kernel</u> K.

Proof: Let $x \in Nu$. Suppose $x \notin K$. Then there exist i and j in $B_k \in B$ such that $s_{ij}(x) > s_{ji}(x)$ and $x^j > v(\{j\})$.

Let p and q be the smallest indices for which

$$s_{ij}(x) = v(S_p) - x(S_p) \quad \text{and} \quad s_{ji}(x) = v(S_q) - x(S_q) \quad .$$

Then $v(S_p) - x(S_p) > v(S_q) - x(S_q)$ (and $i \in S_p$, $j \notin S_p$; $i \notin S_q$, $j \in S_q$). Since $x^j > v(\{j\})$ there exists $\epsilon > 0$ such that y is in X_B where

$$y = (x^1, \ldots, x^{i-1}, x^i + \epsilon, x^{i+1}, \ldots, x^{j-1}, x^j - \epsilon, x^{j+1}, \ldots, x^n) \quad .$$

For $r < p$, $x(S_r) = y(S_r)$ because either both i and j are in S_r or neither i nor j is in S_r. (If not, the condition $s_{ij}(x) = v(S_p) - x(S_p)$ would be contradicted.) Hence $\theta_r(x) = \theta_r(y)$.

For $r = p$, $v(S_p) - y(S_p) = v(S_p) - x(S_p) - \epsilon$. Hence for ϵ sufficiently small $\theta_p(x) > \theta_p(y)$. So, finally, there exists y in X_B such that $\theta(x) \gtrless \theta(y)$. This contradicts the fact that x is in the Nucleolus, and the proof of the theorem is complete.

7.19 Remark: Theorems 7.12, 7.17, and 7.18 imply that the Bargaining Set is nonempty for any coalition structure B (so long as $X_B \neq \emptyset$).

The Nucleolus has many other interesting properties, among which are the following, which will not be proved here.

7.20 Theorem (Schmeidler [1969]): The Nucleolus contains only one point.

7.21 Theorem (Kohlberg [1971] and Schmeidler [1969]): The Nucleolus is a continuous function of v.

7.22   Remark: Neither the Kernel nor the Bargaining Set is necessarily
a continuous function of v (see Stearns [1968]).

7.23   Theorem (Peleg [1968]): For $B = \{N\}$, the Nucleolus of a homogeneous strong weighted majority game consists of the normalized homogeneous weights.

The Kernel has the following interesting property:

7.24   Theorem (Maschler and Peleg [1966]): For $B = \{N\}$, if the Kernel
and the core are nonempty, then the intersection of the Kernel and the core
is nonempty.

Moreover, a point in the Kernel represents an exact compromise within
the core between a pair of players. More precisely, let $x = (x^1,\ldots,x^n)$
be in the core. The set $\{y: y^k = x^k$ for $k \neq i$, $k \neq j$ and $y(N) = x(N)\}$
is a line. The intersection of this set with the core is a segment. A
point belongs to the intersection of the Kernel and the core if and only
if it is the midpoint of all such $\binom{n}{2}$ segments, for some x in the core.

Chapter 8: Repeated Games

When a game is repeated many times, it seems that some sort of
"cooperative" behavior might be induced: if a player deviates from a particular strategy at some point in order to increase his own payoff, the
other players may be able to act in such a way that he is penalized in
every subsequent play of the game. To formalize these ideas, let G be
a game in strategic form (see Definition 2.3). The supergame G^* of G

is then the game each play of which consists of an infinite sequence of plays of G. One might then expect that the outcomes in G^* generated by noncooperative solution concepts (i.e. ones in which it is assumed that contracts cannot be enforced) are related to the outcomes in G generated by cooperative solution concepts (where it is assumed that contracts can be enforced). To examine this question, use the following notations for G:

(1) $N = \{1,\ldots,n\}$ is the set of players,

(2) Σ^i is the (finite) set of strategies of player i; σ^i is an element of Σ^i and $\sigma \equiv (\sigma^1,\sigma^2,\ldots,\sigma^n) \in \underset{i \in N}{\times} \Sigma^i \equiv \Sigma$, and

(3) $h: \Sigma \to E^N$ is the vector of players' payoff functions.

The player set for G^* is also $N = \{1,\ldots,n\}$; the strategy sets and payoff functions are defined below.

8.1 Definition: A pure strategy in G^* (or a pure super-strategy) for player i is a sequence of functions $f_1^i, f_2^i, \ldots$ where $f_k^i: \underbrace{\Sigma \times \Sigma \times \ldots \times \Sigma}_{k-1} \to \Sigma^i$.

8.2 Remark: An assumption implicit in the definition is that at the k-th play each player knows the strategies which were used by the other players in the k - 1 previous plays. This is information which is not necessarily revealed by the outcome at each play of the game, and so the assumption is a strong one. However, it is made merely for convenience here, and weaker assumptions are sufficient to demonstrate many of the results.

Now, one might consider defining mixed strategies in G^* as before. However, a difficulty arises. Consider those pure strategies of player i in G^* which are sequences of constant functions f_k^i. For each k, the set of f_k^i had the cardinality of Σ^i, so that the set of all such pure strategies, being the Cartesian product of denumerably many copies of Σ^i, has the cardinality of the continuum. Hence defining mixed strategies as probability distributions as before would not be straightforward. For this reason we will think of a mixed strategy as a random device for choosing a pure strategy: it will be a random variable -- a function from a sample space into the space of pure strategies. Thus, for each $S \subseteq N$ let Ω^S be a sample space. (This space has to allow sufficient randomization possibilities, which is the case if it is a copy of $([0,1],\mathcal{B},\lambda)$, where $\mathcal{B}$ is the set of Borel subsets of $[0,1]$ and λ is Lebesgue measure; each coalition can then randomize independently of every other coalition.) Ω^S is the lottery which the coalition S .can observe, on which it can base its randomization. Define

$$\Omega^i \equiv \underset{S \ni i}{\times} \Omega^S \quad \text{and} \quad \Omega \equiv \underset{S \subseteq N}{\times} \Omega^S \ .$$

Let $\omega \in \Omega$ and let ω^S be the projection of ω onto Ω^S. Then $\omega^i = (\omega^S)_{S \ni i}$ is the information available to player i.

8.3 Definition: A randomized super-strategy of i is a sequence F^i of functions f_k^i with $f_k^i: \underbrace{\Sigma \times \Sigma \times \ldots \times \Sigma}_{k-1}, \times \Omega^i \to \Sigma^i$.

8.4 Remark: Note that the sample space Ω^i is not indexed by the serial number k of the play. This means that the randomization is based only on the realization ω^i in Ω^i; this allows independent randomization at each play k of the game since $([0,1],\mathcal{B},\lambda)$ is isomorphic to the Cartesian product of denumerably many copies of itself.

Now let an n-tuple $F = (F^1, \ldots, F^n)$ of randomized super-strategies be given. Define a sequence $\underset{\sim}{\sigma}_1, \underset{\sim}{\sigma}_2, \ldots$ of n-tuples of randomized strategie in G as follows:

$$\underset{\sim}{\sigma}_1(\omega) = (f_1^1(\omega^1), f_1^2(\omega^2), \ldots, f_1^n(\omega^n))$$

and

$$\underset{\sim}{\sigma}_k(\omega) = (f_k^1(\underset{\sim}{\sigma}_1(\omega), \ldots, \underset{\sim}{\sigma}_{k-1}(\omega), \omega^1), \ldots, f_k^n(\underset{\sim}{\sigma}_1(\omega), \ldots, \underset{\sim}{\sigma}_{k-1}(\omega), \omega^n))$$

$$\text{for } k = 2, 3, \ldots \ .$$

Define a sequence of random payoffs by

$$\underset{\sim}{h}_k^F(\omega) = h(\underset{\sim}{\sigma}_k(\omega)) \ ,$$

and let

$$H_k(F) = E(\underset{\sim}{h}_k^F) \qquad \text{(where } E \text{ is the expectation operator)} \ .$$

One might then consider taking the expected average payoff $\lim\limits_{m \to \infty} \sum\limits_{k=1}^{m} H_k(F)/m$ as the payoff in the supergame. However, this limit does not always exist. In fact, there is no need to define a payoff function for G^*: we can merely define equilibrium points in the following ways.

8.5 <u>Definition</u>: An n-tuple F_* of randomized super-strategies is an <u>upper equilibrium point</u> in G^* if:

1. $\sum\limits_{k=1}^{m} \underset{\sim}{h}_k^{F_*}(\omega)/m$ converges to a constant $L(F_*)$ with probability one $(L(F_*)$ is referred to as the <u>payoff</u> to the upper e.p.), and

2. for each i and each randomized super-strategy F^i of i there is no $\epsilon > 0$ such that for infinitely many m

$$\frac{1}{m} \sum_{k=1}^{m} (h_k^{F_*|F^i}(\omega))^i > L^i(F_*) + \epsilon \text{ with positive probability } .$$

8.6   Definition: An n-tuple F_* of randomized super-strategies is a lower equilibrium point in G^* if:

1. $\sum_{k=1}^{m} h_k^F{}_*(\omega)/m$ converges to a constant $L(F_*)$ with probability one ($L(F_*)$ is referred to as the payoff to the lower e.p.), and

2. for each i and each randomized super-strategy F^i of i there is no $\epsilon > 0$ such that for all but finitely many m

$$\frac{1}{m} \sum_{k=1}^{m} (h_k^{F_*|F^i}(\omega))^i > L^i(F_*) + \epsilon \text{ with probability one } .$$

8.7   Remark: An upper e.p. is clearly a lower e.p., but the reverse is not necessarily true. However, we will establish below (Theorem 8.14) that the set of payoffs to upper e.p.'s coincides with the set of payoffs to lower e.p.'s.

8.8   Remark: An upper e.p. corresponds to an "optimistic" outlook by each player: F will not be an upper e.p. if player i has a strategy F_0^i for which there is an $\epsilon > 0$ such that for infinitely many m, $\sum_{k=1}^{m} (h_k^{F|F_0^i}(\omega))^i/m$ $> L^i(F_*) + \epsilon$ with positive probability, yet such a deviation might benefit player i only very infrequently. Similarly, a lower e.p. corresponds to a "pessimistic" outlook by each player.

We will now formalize two notions of equilibrium which consider deviations by sets of players.

8.9 **Definition**: An n-tuple F_* of randomized super-strategies is an **upper strong equilibrium point** in G^* if:

1. $\sum_{k=1}^{m} h_k^{F_*}(\omega)/m$ converges to a constant $L(F_*)$ with probability one,

and 2. there is no coalition S, no $|S|$-tuple of randomized super-strategies F^S of S, and no $\varepsilon > 0$ such that for infinitely many m for all i in S

$$\frac{1}{m} \sum_{k=1}^{m} (h_k^{F_*|F^S}(\omega))^i > L^i(F_*) + \varepsilon \quad \text{with positive probability} \quad .$$

8.10 **Definition**: An n-tuple F_* of randomized super-strategies is a **lower strong equilibrium point** in G^* if:

1. $\sum_{k=1}^{m} h_k^{F_*}(\omega)/m$ converges to a constant $L(F_*)$ with probability one,

and 2. there is no coalition S, no $|S|$-tuple of randomized super-strategies F^S of S, and no $\varepsilon > 0$, such that for all but finitely many m for all i in S

$$\frac{1}{m} \sum_{k=1}^{m} (h_k^{F_*|F^S}(\omega))^i > L^i(F_*) + \varepsilon \quad \text{with probability one} \quad .$$

Before examining the equilibrium payoffs in G^*, we will examine the set of feasible payoff vectors in G. Let $\sigma \in \Sigma$; then $h(\sigma) \in E^N$ is the vector of payoffs to players in G. Let

$$P = \{h(\sigma): \sigma \in \Sigma\} \quad \text{and} \quad D = \text{convex hull of } P \quad .$$

Any payoff in D can be attained by the players using jointly mixed strategies.

8.11   Example: Consider the game   G   defined in the table below. We have

$P = \{(1,0),(0,1),(0,0)\}$ and $D = \{x \in E^2: x^1 + x^2 \leq 1, x^1 \geq 0$ and $x^2 \geq 0\}$,

as shown in the diagram. Those points attainable by <u>independent</u> randomiza-

tion comprise the set $A = \{(x^1,x^2): x^1 = \alpha\beta, x^2 = (1 - \alpha)(1 - \beta), \alpha \in [0,1]$

and $\beta \in [0,1]\}$; in order to attain all points in   D   the two players will

	L	R
T	1,0	0,0
B	0,0	0,1

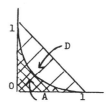

have to correlate their strategies. For example, to attain the payoff vector

$(1/2,1/2)$, they will have to play   T   and   L   together with probability $1/2$,

and   B   and   R   together with probability $1/2$.

8.12   <u>Definition</u>: <u>The minimax payoff to player</u>   i   <u>in</u>   G   <u>is</u>

$$d^i = \min_{\underset{\sim}{\tau}} \max_{\underset{\sim}{\sigma}} E[h^i(\underset{\sim}{\sigma},\underset{\sim}{\tau})] \quad ,$$

where $\underset{\sim}{\sigma}$ (resp. $\underset{\sim}{\tau}$)   runs over the randomized strategies of   i   (resp.

$N\backslash\{i\}$)   in   G.

One can interpret d^i as a payoff which player   i   can guarantee

himself: even if the players in $N\backslash\{i\}$ get together and act so as to make

his payoff as small as possible, he will be able to obtain d^i.

8.13   <u>Example</u>: Consider a three-person game   G   where player 3's payoffs

are as in the tables below. Then in order to minimize player 3's payoff, players 1 and 2 have to randomize between the strategy pairs (T,L) and (B,R): they cannot do so by randomizing independently.

	L	R
T	-1	0
B	0	0

3's first strategy

	L	R
T	0	0
B	0	-1

3's second strategy

8.14 Theorem: The set of payoffs to upper e.p.'s coincides with the set of payoffs to lower e.p.'s and is equal to $D' = \{x \in D: x^i \geq d^i \text{ for all } i \in N\}$

To prove the theorem we need the following.

8.15 Lemma: Consider a two-person zero-sum game G with minimax value v. Let σ be an optimal strategy for player 1 in G, and suppose that in the supergame G^* player 1 uses the randomized super-strategy F^1 which involves the use of an independent copy of σ at each stage. Let F^2 be any randomized super-strategy of player 2 in G^* (in particular, the strategies 2 uses in successive plays of G may not be independent). Then with probability one $\liminf \sum_{k=1}^{m} (h_k^F(\omega))^1/m \geq v$.

Proof: Let the strategy which player 2 uses at stage k be τ_k. Then we know that

$$E((h_k^F)^1 | \sigma_1, \tau_1, \ldots, \sigma_{k-1}, \tau_{k-1}) \geq v \quad \text{for all } k \ .$$

Define

$$x_k(\omega) = (h_k^F(\omega))^1 + v - E((h_k^F)^1 | \sigma_1(\omega), \tau_1(\omega), \ldots, \sigma_{k-1}(\omega), \tau_{k-1}(\omega)) \ .$$

Then $x_k(\omega) \leqq (h_{-k}^F(\omega))^1$ for all k and $E(x_k | \sigma_{-1}, \tau_{-1}, \ldots, \sigma_{k-1}, \tau_{k-1}) = v$

for all k. Also, x_k and x_ℓ are uncorrelated for any k and ℓ (with

$k \neq \ell$): for $k > \ell$

$$E(x_k x_\ell) = E(E(x_k x_\ell | x_\ell)) = E(x_\ell E(x_k | x_\ell)) \quad .$$

But $E(x_k | x_\ell) = v$ since x_ℓ is a function of $\sigma_{-1}, \tau_{-1}, \ldots, \sigma_{k-1}, \tau_{k-1}$. So

$$E(x_k x_\ell) = v E(x_\ell) = v^2 = E(x_k) E(x_\ell) \quad .$$

So we can apply the Strong Law of Large Numbers to deduce that

$$\lim \sum_{k=1}^{m} \frac{x_k}{m} = v \quad \text{with probability one} \quad ,$$

so that, with $x_k(\omega) \leqq (h_{-k}^F(\omega))^1$ for all k,

$$\lim \inf \sum_{k=1}^{m} \frac{(h_{-k}^F(\omega))^1}{m} \geqq v \quad .$$

This establishes the lemma.

Proof of Theorem: The theorem is equivalent to the following three

statements:

(1) {upper e.p.'s} $\subseteq$ {lower e.p.'s},

(2) {payoffs to lower e.p.'s} $\subseteq$ D', and

(3) D' $\subseteq$ {payoffs to upper e.p.'s}.

(1) is immediate.

(2) is the result of the following.

Assertion: {payoffs to lower e.p.'s} $\subset$ D.

Proof: Let F_* be a lower e.p. We have $h_{\underline{k}}^{F}*(\omega) \in D$ for all k,

so $\sum_{k=1}^{m} h_{\underline{k}}^{F}*(\omega)/m \in D$ for $m = 1,2,\ldots$. Hence the closedness of D ensures

that $L(F_*) \in D$. This proves the assertion.

Assertion: If x is a payoff to a lower e.p. then $x^i \geqq d^i$ for

all $i \in N$.

Proof: Let F_* be a lower e.p. with payoff x. By the definition

of d^i, i has a strategy σ^i such that for all strategies $\tau_{\underline{k}}^{N\setminus\{i\}}$ at stage

k of the supergame, $E[h^i(\tau_{\underline{k}}^{N\setminus\{i\}},\sigma^i)] \geqq d^i$. So if at each stage player i

uses an independent copy of σ^i, Lemma 8.15 guarantees that $L^i(F_*) = x^i \geqq d^i$,

which establishes the assertion.

These two assertions together entail the truth of (2).

(3) can be proved as follows. Let $x \in D'$. Then there exist non-

negative real numbers α_j summing to one and members ξ_j of Σ such that

$\sum_j \alpha_j h(\xi_j) = x$ and $x^i \geqq d^i$ for all $i \in N$. Let $\sigma^N(\omega^N)$ be the mixed

strategy in G which involves the players using the pure strategy ξ_j with

probability α_j, for all j, so that $E[h(\sigma^N(\omega^N))] = x$. Define F_*^i by

$$f_{*k}^i(\sigma_1,\sigma_2,\ldots,\sigma_{k-1},\omega^i) = \begin{cases} (\sigma_{\underline{k}}^N(\omega^N))^i & \text{if } \sigma_{\underline{\ell}} = \sigma_{\underline{\ell}}^N(\omega^N) \text{ for all } \ell \leqq k-1 \\ \tau_{\underline{i}'}^i,(\omega^{N\setminus\{i'\}}) & \text{otherwise} \end{cases},$$

where the $\sigma_{\underline{\ell}}^N(\omega^N)$ are independent copies of $\sigma^N(\omega^N)$, i' is the first player

to deviate from the strategy $\sigma^N(\omega^N)$ in the previous plays (first in time;

if there are more than one such, first in serial number), and $\tau_{i'}$ is a correlated strategy of $N\setminus\{i'\}$ with $\max_{\sigma} E[h^{i'}(\sigma,\tau_{i'})] = d^{i'}$ (σ running over randomized strategies of i' in $\tilde{G}$). Suppose i' deviates from F_* at stage ℓ, and uses the strategies $\hat{\sigma}_k^{i'}$ for $k = \ell, \ell+1, \ldots$; let $F^{i'}$ denote his super-strategy. Then

$$E[h^{i'}(\hat{\sigma}_k^{i'}, (\tau_{i'})_k)] \leq d^i \quad \text{for all} \quad k \geq \ell+1 \quad ,$$

where the $(\tau_{i'})_k$ are independent copies of $\tau_{i'}$. So by Lemma 8.15

$$\limsup_{k=\ell+1}^{m} (h_k^{F_*|F^{i'}}(\omega))^{i'}/(m-\ell) \leq d^{i'} \quad \text{with probability one. Hence}$$

for i' condition 2 of Definition 8.5 is met. Hence no player can gain by deviating from the strategy $\sigma^N(\omega^N)$, and so F_* is an upper e.p. with payoff x. This establishes (3).

This completes the proof of the theorem.

8.16 Example: Consider the Prisoners' Dilemma game, with payoff matrix as below (see Example 2.20). The set D is the parallelogram with vertices consisting of the four payoff vectors in the table; D' is the subset of D with $(x^1, x^2) \geq (1,1)$, since 1 is the minimax payoff of both players.

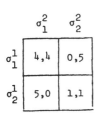

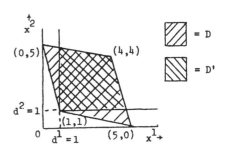

Theorem 8.14 states that the set of payoffs to e.p.'s in the supergame consisting of an infinite string of repetitions of the Prisoners' Dilemma game consists precisely of D'. This contrasts with the equilibrium payoff vector in the Prisoners' Dilemma game when it is played once, which is just (1,1).

Now let

$$v_\alpha(S) = \{x^S \in E^S: \text{ there exists a randomized strategy } \sigma^S \text{ of } S$$
$$\text{in } G \text{ such that for all randomized strategies } \underset{\sim}{\sigma}^{N\setminus S} \text{ of}$$
$$N\setminus S \text{ in } G \text{ and for all } i \text{ in } S, x^i \leq E[h^i(\underset{\sim}{\sigma}^S, \underset{\sim}{\sigma}^{N\setminus S})]\}.$$

This is just the set of all payoff vectors which S can guarantee for itself.

8.17 <u>Definition</u>: The α-<u>core of</u> G is the core of v_α (i.e. α-core of $G = \{x \in D: \ (\forall S \subset N)(\not\exists y \in v_\alpha(S) \text{ s.t. } y^i > x^i \ \forall i \in S)\})$.

Next, let

$$v_\beta(S) = \{x^S \in E^S: \text{ for each randomized } \underset{\sim}{\sigma}^{N\setminus S} \text{ of } N\setminus S \text{ in } G, \text{ there}$$
$$\text{exists a randomized strategy } \underset{\sim}{\sigma}^S \text{ of } S \text{ in } G \text{ such that}$$
$$\text{for all } i \text{ in } S, x^i \leq E[h^i(\underset{\sim}{\sigma}^S, \underset{\sim}{\sigma}^{N\setminus S})]\}.$$

This is the set of all payoff vectors that N\S cannot prevent S from getting.

8.18 <u>Definition</u>: The β-<u>core of</u> G is the core of v_β (i.e. β-core of $G = \{x \in D: \ (\forall S \subset N)(\not\exists y \in v_\beta(S) \text{ s.t. } y^i > x^i \ \forall i \in S)\})$.

8.19   Remark: $v_\alpha(S) \subset v_\beta(S)$, so ($\beta$-core of v) $\subset$ (α-core of v).

It would seem that v_α is a more natural construct than v_β, but in connection with repeated games, it is v_β that turns out to be more significant. Thus, we have the following.

8.20   <u>Theorem</u> (Aumann [1959]): <u>The set of payoffs to upper strong e.p.'s in G^* coincides with the set of payoffs to lower strong e.p.'s in G^*, and with the β-core of</u> G.

Proof: From the definitions we can immediately deduce that

{upper strong equilibrium payoffs} $\subset$ {lower strong equilibrium payoffs}

The theorem is then equivalent to the following two statements:

(1)   {lower strong equilibrium payoffs} $\subset$ β-core, and

(2) β-core $\subset$ {upper strong equilibrium payoffs}.

To establish (1), suppose that F_* is a lower strong e.p. with payoff x, and $x \notin \beta$-core. To say that $x \notin \beta$-core means that there exists a coalition S and $y \in v_\beta(S)$ with $y^i > x^i$ for all $i \in S$; $y \in v_\beta(S)$ means that

for each randomized strategy $\tau^{N \backslash S}$ of $N \backslash S$ in G there exists a randomized strategy σ^S of S in G with $E[h^i(\sigma^S, \tau^{N \backslash S})] \geqq y^i$ for all $i \in S$.

So for each sequence of n-tuples of strategies $\sigma_1, \ldots, \sigma_{k-1}$, there exists a randomized strategy $\sigma_k^S: \Omega^S \to \Sigma^S$ for S in play k of the game such that for all $i \in S$

$$E[h^i(\sigma^S_{-k}(\omega^S), f^{N\backslash S}_{*k}(\sigma_{-1}, \ldots, \sigma_{-k-1}, \omega))] \geqq y^i$$

Now define supergame strategies F^i for $i \in S$ as follows:

(a) $f^i_1(\omega) = \sigma^i_{-1}(\omega^S)$;

(b) $f^i_k(\sigma_{-1}, \sigma_{-2}, \ldots, \sigma_{-k-1}, \omega) = \sigma^i_{-k}(\omega^S)$ for $k = 2, 3, \ldots$.

Then $E[(h^{F_*|F^S}_{-k}(\omega))^i] \geqq y^i$ for all $i \in S$ and for all stages k. But $y^i > x^i$ for all $i \in S$, so Lemma 8.15 ensures that for all but finitely many m, for some $\varepsilon > 0$

$$\frac{1}{m} \sum_{k=1}^{m} (h^{F_*|F^S}_{-k}(\omega))^i > x^i + \varepsilon \quad \text{with probability one}$$

contradicting the fact that F_* is a lower strong equilibrium payoff. This establishes (1).

In order to prove (2), we will use the following.

Assertion: For each $x \in \beta$-core and each $S \subset N$, there exists a randomized strategy $\tau^{N\backslash S}$ of $N\backslash S$ in G such that for each randomized strategy σ^S of S in G there exists an $i \in S$ such that $E[h^i(\sigma^S, \tau^{N\backslash S})] \leqq x^i$.

Proof: Since $x \in \beta$-core, there is no $y \in v_\beta(S)$ such that $y^i > x^i$ for all $i \in S$. So for all $\varepsilon > 0$, there is a strategy $\tau^{N\backslash S}_\varepsilon$ of $N\backslash S$ such that for each strategy σ^S of S there is an $i \in S$ with $E[h^i(\sigma^S, \tau^{N\backslash S}_\varepsilon)] < x^i + \varepsilon$. The i for which this is true will depend on ε, but as ε tends to zero through a denumerable sequence there will be an i such that the statement is true for infinitely many terms of the sequence, and for this i we will have inequality in the limit, establishing the assertion.

Now we can prove (2). Let $x \in \beta$-core. Let $\sigma^N(\omega^N)$ be a strategy of N in G for which $E[h(\sigma^N(\omega^N))] = x$. By the assertion, for each $S \subset N$ there exists a randomized strategy $\tau^{N \backslash S}$ of $N \backslash S$ in G such that for each randomized strategy σ^S of S in G there is an $i \in S$ such that $E[h^i(\sigma^S, \tau^{N \backslash S})] \leq x^i$. We will use the strategies σ^N and $\tau^{N \backslash S}$ to define a randomized super-strategy F_* as follows:

(a) $f_{*1}(\omega) = \sigma_1^N(\omega^N)$, a copy of $\sigma^N(\omega^N)$;

(b) let S_k be the set of players who have used a strategy other than their component of σ^N in one or more of the plays $1, \ldots, k-1$. Then

$$f_{*k}^{N \backslash S_k}(\sigma_1, \ldots, \sigma_{k-1}, \omega) = \tau_{-k}^{N \backslash S_k}(\omega^{N \backslash S_k}) \quad \text{if } S_k \neq \emptyset$$

(where $\tau_{-k}^{N \backslash S_k}$ is a copy of $\tau^{N \backslash S_k}$, independent of $\tau_{-j}^{N \backslash S_j}$ for $j \leq k-1$), and

$$f_{*k}^N(\sigma_1, \ldots, \sigma_{k-1}, \omega) = \sigma_k^N(\omega^N) \quad \text{if } S = \emptyset$$

(where the σ_k^N are independent copies of σ^N). Suppose that after play $\ell - 1$ no new players deviate from σ^N; let $S_\ell = S$, and let F^S be the randomized super-strategy of S, ρ_{-k}^S being the strategy S uses at play k for $k \geq \ell$.

Let C be the set of $|S|$-tuples of expected payoffs to members of S in G which S can attain when $N \backslash S$ uses the strategy $\tau^{N \backslash S}$. C is a convex subset of E^S, and x is not in the interior of C (if it were, there would exist a strategy σ^S of S such that $E[h^i(\sigma^S, \tau^{N \backslash S})] > x^i$ for all $i \in S$). So by the Supporting Hyperplane Theorem there exists $p \in E^S$, $p \neq 0$, such that

$$\sum_{i \in S} p^i y^i \leq \sum_{i \in S} p^i x^i \quad \text{for all } y \in C \ .$$

Since there is no $z \in C$ such that $z \gg x$, we can deduce that there is a supporting p with $p > 0$; i.e. there exists $p > 0$ such that

$$\sum_{i \in S} p^i E[h^i(\rho_{-k}^S, \tau_{-k}^{N \backslash S})] \leq \sum_{i \in S} p^i x^i \quad \text{for all strategies } \rho_{-k}^S \text{ of } S$$
$$\text{for all } k \geq \ell \ .$$

Hence by Lemma 8.15,

$$\limsup_{} \sum_{k=\ell}^{m} \sum_{i \in S} p^i \frac{h^i(\rho_{-k}^S, \tau_{-k}^{N \backslash S})}{m - \ell} \leq \sum_{i \in S} p^i x^i \ .$$

So there exists no $\epsilon > 0$ such that for infinitely many m for all $i \in S$

$$\sum_{k=1}^{m} \frac{(h_{-k}^{F_*|F^S}(\omega))^i}{m} > x^i + \epsilon \quad \text{with positive probability}$$

Hence F_* is an upper strong equilibrium point with payoff x.

In this argument we have assumed that S is a constant. In fact it may be a random variable (whether a player deviates or not may depend on the strategies the other players have used in previous stages), so that the p above will be random, and the expectations will all be conditional. The line of argument needs no modification, however, and we leave it to the reader to make the necessary notational changes.

This completes the proof of (2), so that the theorem is established.

8.21   Example: Consider the Prisoners' Dilemma game once again, with payoff matrix as given in Example 8.16. We find that the β-core of this game consists of the line segments ab and bc shown in the diagram, so by

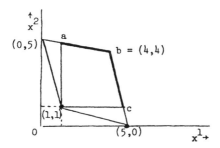

Theorem 8.20 the points on these lines are the strong equilibrium payoffs in the supergame, an outcome which contrasts once again with the noncooperative outcomes in the game when it is played once.

Appendix to Chapter 8: Annotated Bibliography on Repeated Games

(a) Repeated Games and Cooperation:

Three   papers in this area are Aumann [1959], [1961], and [1967]. The first introduces the material examined above, but analyzes mixed strategies in the supergame as probability distributions rather than random variables, and is consequently difficult to read. The second analyzes the

α-core and the β-core in the case where side payments are not allowed; the
worth of a coalition is then the set of payoff vectors to its members
which it can attain, rather than a single amount which can be distributed
within the coalition in any way. The third paper surveys many of the
topics discussed in these lectures for games without side payments. An
application of the idea that repetition leads to outcomes reflecting coop-
eration is contained in Kurz [1977].

(b) Stochastic Games:

Stochastic games were the first sort of repeated games to be analyzed.
A stochastic game is a finite set of 2-person zero-sum games, each play of
the game leading to a payoff and the assignment of some game in the set, the
latter being played at the next stage. So each player can maneuver for payoffs and for subsequent games. Shapley [1953b] analyzes the case where the
payoffs are discounted, and Gillette [1957] examines the case where they are
not. Since then a great deal of work has been devoted to stochastic games;
it is so voluminous that it cannot possibly all be reviewed here. Dramatic
progress has recently been made by Bewley and Kohlberg [1976a] and [1976b]
in studying the undiscounted case.

(c) Repeated Games with Incomplete Information:

In these games one game out of a known set is played repeatedly, each
player having only limited information about which game is being played.
Each player will then be interested not only in obtaining a high payoff in
the short term, but in ensuring that he can do so over a long period by
playing so as to conceal any information he has which the other players do

not have (if players' interests are opposed) or by playing so as to reveal

information (if his interests coincide with those of others).

As an example of a situation where interests are opposed, assume

the set of games consists of two zero-sum two-person games with payoffs to

player 1 as shown in the tables below. Player 1 (the row player) knows

Game 1   Game 2

which game is being played, but player 2 does not (we assume that player 2

does not know the payoff he receives at the end of any play; this payoff is

just deposited into his bank account). If player 1 always plays his top

strategy, it will be clear to player 2 that game 1 is being played, so the

best strategy for player 1 will involve his playing his bottom strategy some

of the time, in order to avoid revealing his information to player 2.

Now consider a situation where players' interests coincide. The

games in the set are the two whose payoff matrices are shown below. Player

1 (the row player) has one strategy, and player 2 has two. Assume that

player 1 knows which game is being played, but player 2 does not. Even

though players' interests coincide, the outcome will not be efficient in

Game 1   Game 2

this case since the fact that player 1 has only one strategy means that he cannot signal to player 2 the information he has about the game being played; there is no way for the players to coordinate their actions. If, however, player 1 has two strategies, the two games being Game 3 and Game 4, whose payoff matrices are shown below, then there is some possibility for coordination: whenever the true game is Game 3 (resp. 4) player 1 can play his top (resp. bottom) strategy, and if he does so, player 2 can play his left (resp. right) hand strategy. The resulting equilibrium point will be efficient.

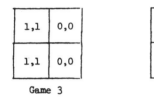

1,1	0,0
1,1	0,0

Game 3

0,0	1,1
0,0	1,1

Game 4

The same outcome would result if player 1 could signal to player 2 in some other way. But if the payoffs are modified slightly a new problem arises. Thus consider the case where the games are Games 5 and 6. A situation where

0,1	0,0
1,1	0,0

Game 5

1.1,0	1,1
1.1,0	1,1

Game 6

player 2 plays his left hand strategy if player 1 signals that game 5 is being played will not be sustainable as an equilibrium point in this case because there will then always be an incentive for player 1 to signal that game 5 is being played when in fact game 6 is being played. In this case, then, there is no efficient equilibrium point even though there is the possibility of signalling.

Some papers which analyze repeated games with incomplete information are Aumann and Maschler [1966], [1967], and [1968], Stearns [1967], Kohlberg [1975a] and [1975b], Mertens and Zamir [1971/72], and Zamir [1971/72] and [1973]. Again, the literature is too voluminous to review completely here.

Chapter 9: Some Final Remarks

A topic which we have not covered is <u>utility theory</u>. When we introduced the payoff matrix in two-person games the numbers we assigned were only intended to represent players' orderings over the possible outcomes. However, subsequently it was necessary to interpret the payoffs as representing preference intensities. Thus, when considering the mixed extension of a game we dealt with expected payoffs, which are sums of payoffs weighted by probabilities, and when we analyzed cooperative games we assumed there existed numbers representing the "worth" of each coalition. We can justify such procedures by assuming that payoffs are in money units and that each player has a utility function which is linear in terms of money. For a detailed treatment of the problem see Luce and Raiffa [1957].

We have also not studied some other interesting topics. Zermelo's theorem involves a <u>game in extensive form</u>, where the sequential structure of the players' moves is considered in detail. <u>Games in coalitional form</u>

without side payments are also of interest, as are games with a continuum of players; the latter can be used to formalize economic situations of pure competition.

Solutions to Exercises[1]

1. Consider the game defined in the table below. Let α_i (resp. β_i) be the probability that player 1 (resp. 2) uses strategy i. Then the only e.p. of the game is $(\alpha_1, \alpha_2, \alpha_3; \beta_1, \beta_2, \beta_3) \doteq (1/3, 1/3, 1/3; 1/3, 1/3, 1/3)$, the equilibrium payoff being $(3,3)$.

	s_1^2	s_2^2	s_3^2
s_1^1	0,0	4,5	5,4
s_2^1	5,4	0,0	4,5
s_3^1	4,5	5,4	0,0

Proof: By inspection there is no pure strategy e.p. Let $(\alpha_1, \alpha_2, \alpha_3; \beta_1, \beta_2, \beta_3)$ be a mixed strategy e.p. Then given $(\alpha_1, \alpha_2, \alpha_3)$ it must be the case that the expected payoff to player 2 if he uses any of his pure strategies is the same (otherwise he would choose a pure strategy, in which case it would also be best for player 1 to choose a pure strategy, so that we would not have a mixed strategy e.p.). Hence we need

$$4\alpha_2 + 5\alpha_3 = 5\alpha_1 + 4\alpha_3 = 4\alpha_1 + 5\alpha_2 \quad ,$$

---

[1] Supplied by Martin J. Osborne.

or, remembering that $\alpha_1 + \alpha_2 + \alpha_3 = 1$, $\alpha_1 = \alpha_2 = \alpha_3 = 1/3$. By symmetry we can then deduce that $\beta_1 = \beta_2 = \beta_3 = 1/3$. The payoff to player 1 at this e.p. is $4(1/3) + 5(1/3) = 3$, as is the payoff to player 2. Hence the only e.p. of the game is $(1/3,1/3,1/3; 1/3,1/3,1/3)$, the equilibrium payoff being $(3,3)$.

2.   Consider the game defined in the tables below. Let α (resp. β, γ) be the probability that   player 1 (resp. 2, 3) uses his first strategy. Then the e.p.'s of this game are $(\alpha,\beta,\gamma) = (1,1,1)$, $(\alpha,\beta,\gamma) = (0,0,0)$, and $(\alpha,\beta,\gamma) = (2-\sqrt{2},2-\sqrt{2},2-\sqrt{2})$. The corresponding equilibrium payoffs are $(1,1,1)$, $(2,2,2)$, and $(\sqrt{2},\sqrt{2},\sqrt{2})$.

Proof: $(\alpha,\beta,\gamma) = (1,1,1)$ and $(\alpha,\beta,\gamma) = (0,0,0)$ are clearly pure strategy e.p.'s, with payoffs $(1,1,1)$ and $(2,2,2)$. To find mixed strategy e.p.'s, let players 1 and 2 choose probabilities α and β. Then in a mixed strategy e.p. player 3 must get the same payoff when he uses either of his pure strategies (otherwise he would choose a pure strategy, which

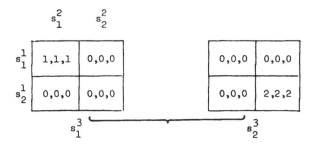

would induce players 1 and 2 to do the same, and we would have a pure strategy e.p.). Hence we want

$$\alpha\beta = 2(1 - \alpha)(1 - \beta)$$

By symmetry the corresponding conditions when the other players are considered are

$$\alpha\gamma = 2(1 - \gamma)(1 - \alpha) \quad \text{and} \quad \beta\gamma = 2(1 - \beta)(1 - \gamma)$$

The unique solution $(\alpha,\beta,\gamma) \in [0,1]^3$ of these three equations is (α,β,γ) = $(2-\sqrt{2}, 2-\sqrt{2}, 2-\sqrt{2})$, and the corresponding equilibrium payoff is $(\sqrt{2},\sqrt{2},\sqrt{2})$. Thus the e.p.'s and equilibrium payoffs are precisely those stated above.

3.   Consider the game defined in the table below. Let α (resp. β) be the probability that player 1 (resp. 2) plays his first strategy. Then every (α,β) in $[0,1]^2$ is an e.p.; the equilibrium payoff corresponding to (α,β) is $(1 - \beta, 1 - \alpha)$.

Proof: Using the arguments of Exercises 1 and 2 it is immediate that the above assertion is correct.

	s_1^2	s_2^2
s_1^1	0,0	1,0
s_2^1	0,1	1,1

4.   If $1 \geq [v(12) + v(13) + v(23)]/2$ and $v(ij) \leq 1$ for all pairs $\{i,j\} \subset \{1,2,3\}$, then the   0-1   normalized 3-person game $(\{1,2,3\},v)$ has a nonempty core.

Proof: (a) Suppose that $v(12) + v(13) \geq 1$. Then

$$x = (v(12) + v(13) - 1, 1 - v(13), 1 - v(12))$$

is a member of the core: $x^1 \geq 0$, $x^2 \geq 0$, $x^3 \geq 0$ and $x^1 + x^2 + x^3 = 1$; and $x^1 + x^2 = v(12)$, $x^1 + x^3 = v(13)$, and $x^2 + x^3 = 2 - v(12) - v(13) \geq v(23)$.

(b) Suppose that $v(12) + v(13) < 1$. Then

$$x = (0, 1 - v(13), v(13))$$

is a member of the core: $x^1 \geq 0$, $x^2 \geq 0$, $x^3 \geq 0$ and $x^1 + x^2 + x^3 = 1$; and $x^1 + x^2 = 1 - v(13) > v(12)$, $x^1 + x^3 = v(13)$ and $x^2 + x^3 = 1 \geq v(23)$.

Hence in all cases the core is nonempty.

5. A 0-1 normalized weighted majority game has a nonempty core if and only if it has at least one veto player.

Proof: A 0-1 normalized weighted majority game (N,v) is defined by

$$v(S) = \begin{cases} 1 & \text{if} \quad \sum_{i \in S} w^i \geq q \\ 0 & \text{if} \quad \sum_{i \in S} w^i < q, \end{cases}$$

with q such that $v(i) = 0$ for all $i \in N$ and $v(N) = 1$.

(a) Sufficiency: Number the players in such a way that $w^1 \geq w^2 \geq \cdots \geq w^n$. Then if there is at least one veto player, player 1 is such a player (i.e. $v(S) = 0$ if $1 \notin S$). Hence $x = (1,0,0,\ldots,0)$ is in the core: $\sum_{i \in N} x^i = 1$, $x^i \geq 0$ for all i, and

$$\sum_{i \in S} x^i = \begin{cases} 1 & \text{if } 1 \in S \text{ , in which case } v(S) \leq 1 \\ 0 & \text{if } 1 \notin S \text{ , in which case } v(S) = 0 \end{cases} \text{ .}$$

This establishes sufficiency.

(b) <u>Necessity</u>: Suppose there are no veto players. Consider the collection of coalitions $S = \{N\backslash\{1\}, N\backslash\{2\},\ldots,N\backslash\{n\}\}$ with balancing weights $\delta_S = 1/(n-1)$ for all $S \in S$. We have

$$\sum_{S \in S} \delta_S x_S = \frac{1}{n-1} \sum_{S \in S} x_S = \frac{1}{n-1} \begin{pmatrix} n-1 \\ n-1 \\ \vdots \\ n-1 \end{pmatrix} = x_N \text{ ,}$$

so S is a balanced collection. Since there are no veto players $v(S) = 1$ for all $S \in S$, so $\sum_{S \in S} \delta_S v(S) = n/(n-1) > 1 = v(N)$. Hence by the Bondareva-Shapley theorem the core of (N,v) is empty. This establishes necessity.

6.   The core of a   0-1   normalized weighted majority game with veto players $1,\ldots,p$ is $C = \{x: x = (a^1,a^2,\ldots,a^p,0,0,\ldots,0)$ with $a^i \geq 0$ for all i and $\sum_{i=1}^{p} a^i = 1\}$.

Proof: If $x \in C$, then it is clearly an imputation. Let S be such that $v(S) = 1$; then $\{1,2,\ldots,p\} \subseteq S$, so $\sum_{i \in S} x^i = 1 = v(S)$. Let S be such that $v(S) = 0$. Then $\sum_{i \in S} x^i \geq 0 = v(S)$. Hence any $x = (a^1,a^2,\ldots,a^p,0,0,\ldots,0)$ with $a^i \geq 0$ for all i and $\sum_{i=1}^{p} a^i = 1$ is in the core of the game, and the core consists of solely such points.

7.   For every concave function $f: E^n \to \mathbb{R}$ and for all $m \geq 1$,
$$f\left(\sum_{i=1}^{m} \alpha^i x^i\right) \geq \sum_{i=1}^{m} \alpha^i f(x^i) \text{ if } \alpha \in E_+^m \text{ and } \sum_{i=1}^{m} \alpha^i = 1.$$

<u>Proof</u>: We proceed by induction on m.

(a) For $m = 1$ the result is immediate.

(b) Assume the result is true for all $m \leqq r - 1$, and take $\alpha \in E_+^r$ with $\sum\limits_{i=1}^{r} \alpha^i = 1$. Then

$$f(\sum_{i=1}^{r} \alpha^i x^i) = f(\sum_{i=1}^{r-1} \alpha^i x^i + \alpha^r x^r)$$

$$= f[(1 - \alpha^r) \sum_{i=1}^{r-1} (\frac{\alpha^i}{1 - \alpha^r}) x^i + \alpha^r x^r]$$

$$\geqq (1 - \alpha^r) f[\sum_{i=1}^{r-1} (\frac{\alpha^i x^i}{1 - \alpha^r})] + \alpha^r f(x^r)$$

(by the concavity of f) .

But $\sum\limits_{i=1}^{r-1} \alpha^i/(1 - \alpha^r) = 1$, and $\alpha^i/(1 - \alpha^r) \geqq 0$ for $i = 1,\ldots,r - 1$, so that the truth of the result for $m = r - 1$ ensures that

$$f(\sum_{i=1}^{r-1} (\frac{\alpha^i x^i}{1 - \alpha^r})) \geqq \sum_{i=1}^{r-1} (\frac{\alpha^i f(x^i)}{1 - \alpha^r}) .$$

Hence $f(\sum\limits_{i=1}^{r-1} \alpha^i x^i) \geqq \sum\limits_{i=1}^{r-1} \alpha^i f(x^i) + \alpha^r f(x^r) = \sum\limits_{i=1}^{r} \alpha^i f(x^i)$. Hence the result is true for $m = r$.

(a) and (b) together entail the verity of the result for all $m \geqq 1$.

8. Lemma 6.3 is false without the assumption of superadditivity.

<u>Proof</u>: Consider the following game (N,v):

$$N = \{1,2,3\} , \quad v(1) = 2 , \quad v(2) = v(3) = 0 ,$$

$$v(12) = v(13) = v(23) = 3 , \quad \text{and} \quad v(123) = 4 .$$

If x is to be in the core of (N,v) then $x^1 \geq v(1) = 2$, $x^2 + x^3 \geq v(23) = 3$, and $x^1 + x^2 + x^3 = 4$; hence the core is empty. But the imputation $(2,1,1)$ is not dominated by any imputation: the only coalition which can dominate it is $\{2,3\}$, but in any imputation $x^1 \geq 2$, so it is not dominated by any imputation. This establishes the claim.

9.   There is no homogeneous representation of the weighted majority game one of whose representations is $[5; 2,2,2,1,1,1]$.

    Proof: We want to find a representation $[q; w^1,\ldots,w^6]$ such that $\sum_{i \in S} w^i = q$ for all minimal winning coalitions S. That is, we want

(1) $w^1 + w^2 + w^3 = q$

and

(2) $w^i + w^j + w^k = q$

where $(i,j) \in \{(1,2),(2,3),(3,1)\}$ and $k \in \{4,5,6\}$. Adding the equations in (2) we obtain $2(w^1 + w^2 + w^3) + 3w^k = 3q$ for $k = 4,5,6$. Using (1) this gives $3w^k = q$ for $k = 4,5,6$, so that $w^k = q/3$ for $k = 4,5,6$. But then $w^4 + w^5 + w^6 = q$, so that $\{4,5,6\}$ is winning, which is not the case in the original game. This establishes that no homogeneous representation of the game exists.

10. $M(v,\{1234\})$ for the weighted majority game $[3; 2,1,1,1]$ is $\{(\alpha,(1 - \alpha)/3, (1 - \alpha)/3, (1 - \alpha)/3\}_{2/5 \leq \alpha \leq 4/7}$.

Proof: Let $x = (\alpha, \beta, \gamma, \delta)$ be a payoff vector. There are three sorts of objections to consider: (a) those among players 2, 3 and 4; (b) those between players 2, 3 and 4 and player 1; and (c) those between player 1 and one of the players 2, 3 and 4.

(a) Let an objection of 2 against 3 be of the form $(1-\beta-\epsilon, \beta+\epsilon, 0, 0)$. Such an objection exists if $1-\beta-\epsilon \geq \alpha$ for some $\epsilon > 0$, or if $\alpha+\beta < 1$. Player 3 can counterobject with $(1-\gamma, 0, \gamma, 0)$ if $1-\gamma \geq 1-\beta-\epsilon$ for all $\epsilon > 0$, or if $\beta \geq \gamma$. So there is no justified objection of 2 against 3 if

(1) either $\alpha + \beta = 1$, or $\beta \geq \gamma$.

The consideration of objections of 2 against 4, 3 against 2, 3 against 4, 4 against 2, and 4 against 3 leads to the conditions

(2) either $\alpha + \beta = 1$, or $\beta \geq \delta$,

(3) either $\alpha + \gamma = 1$, or $\gamma \geq \beta$ and $\gamma \geq \delta$, and

(4) either $\alpha + \delta = 1$, or $\delta \geq \beta$ and $\delta \geq \gamma$,

for there to be no justified objection.

(b) Consider an objection of 2 against 1 of the form $(0, \beta+\epsilon, (1-\beta-\epsilon)/2, (1-\beta-\epsilon)/2)$; such an objection exists if $1-\beta-\epsilon \geq 2\gamma$ for some $\epsilon > 0$ and $1-\beta-\epsilon \geq 2\delta$ for some $\epsilon > 0$, or if $2\gamma < 1-\beta$ and $2\delta < 1-\beta$. Player 1 can counterobject with $(\alpha, 0, 1-\alpha, 0)$ if $1-\alpha \geq (1-\beta-\epsilon)/2$ for all $\epsilon > 0$, or if $2\alpha \leq 1+\beta$. Hence there is no justified objection to x if

(5) $1-\beta \leq 2\gamma$ or $1-\beta \leq 2\delta$ or $2\alpha \leq 1-\beta$.

Consideration of objections of 3 and 4 against 1 leads to the conditions

(6) $1 - \gamma \leq 2\beta$ <u>or</u> $1 - \gamma \leq 2\delta$ <u>or</u> $2\alpha \leq 1 - \gamma$, and

(7) $1 - \delta \leq 2\beta$ <u>or</u> $1 - \delta \leq 2\gamma$ <u>or</u> $2\alpha \leq 1 - \delta$.

Next let $x = (1,0,0,0)$. Then (5) is violated, so x is not in the
Bargaining Set. Also, if $x = (\alpha, 1 - \alpha, 0, 0)$ with $\alpha < 1$, (2) is violated. Hence
$\alpha + \beta < 1$, and by symmetry we can deduce that $\alpha + \gamma < 1$ and $\alpha + \delta < 1$. So
from conditions (1)-(4) we conclude that $\beta = \gamma = \delta$. So any point in the
Bargaining Set is of the form $(\alpha, (1 - \alpha)/3, (1 - \alpha)/3, (1 - \alpha)/3)$ with $\alpha < 1$.

(c) Consider an objection of 1 against 2 of the form $(\alpha + \epsilon, 0, 1 - \alpha - \epsilon, 0)$;
such an objection always exists since $\beta = (1 - \alpha)/3$. Player 2 can counter-
object with $(0, (1 - \alpha)/3, 1 - \alpha - \epsilon, (1 - \alpha)/3)$ if $5(1 - \alpha)/3 - \epsilon \leq 1$ for all
$\epsilon > 0$, or if $\alpha \geq 2/5$. The consideration of objections of 1 against 3 and 4 by
symmetry yields the same conclusion.

Hence $\beta = (1 - \alpha)/3 \leq 1/5$, so the first two conditions in (5) are not
satisfied; for the last one to be satisfied we need $2\alpha \leq 1 + (1 - \alpha)/3$ or
$\alpha \leq 4/7$, and conditions (6) and (7) lead to the same result.

This exhausts all possibilities for objections, so the Bargaining Set
is $(\alpha, (1 - \alpha)/3, (1 - \alpha)/3, (1 - \alpha)/3)_{2/5 \leq \alpha \leq 4/7}$, as was to be shown.

11. $M(v, \{123\})$ for the game defined in Example 7.1 is $\{(2/3, 1/6, 1/6)\}$.

<u>Proof</u>: Consider a payoff vector $x = (\alpha, \beta, \gamma)$. There are three sorts
of objections to examine: (a) 2 against 3, and 3 against 2; (b) 2 and 3
against 1; and (c) 1 against 2 and 3.

(a) Consider an objection of 2 against 3 of the form $(1 - \beta - \epsilon, \beta + \epsilon, 0)$;
such an objection exists unless $\alpha + \beta = 1$. Player 3 can counterobject with

$(1-\gamma,0,\gamma)$ if $1-\gamma \geq 1-\beta-\epsilon$, or if $\gamma \leq \beta$. So there is no justified objection of 2 against 3 if

(1) <u>either</u> $\alpha + \beta = 1$ <u>or</u> $\gamma \leq \beta$.

By symmetry there is no justified objection of 3 against 2 if

(2) <u>either</u> $\alpha + \beta = 1$ <u>or</u> $\beta \leq \gamma$.

(b) Consider an objection of 2 against 1 of the form $(0,\beta+\epsilon,\frac{1}{2}-\beta-\epsilon)$; such an objection exists if $\frac{1}{2}-\beta-\epsilon \geq \gamma$, or if $\beta+\gamma < 1/2$. So no objection exists if $\beta+\gamma \geq 1/2$. Player 1 can counterobject with $(\alpha,0,1-\alpha)$ if $1-\alpha \geq \frac{1}{2}-\beta-\epsilon$, or if $\alpha \leq \beta+\frac{1}{2}$. Hence there is no justified objection of 2 against 1 if

(3) <u>either</u> $\beta + \gamma \geq \frac{1}{2}$ <u>or</u> $\alpha \leq \beta + \frac{1}{2}$.

By symmetry there is no justified objection of 3 against 1 if

(4) <u>either</u> $\beta + \gamma \geq \frac{1}{2}$ <u>or</u> $\alpha \leq \gamma + \frac{1}{2}$.

(c) Finally, consider an objection of 1 against 2 of the form $(\alpha+\epsilon,0,1-\alpha-\epsilon)$. Such an objection exists if $1-\alpha-\epsilon \geq \gamma$, so there can be no objection only if $\gamma+\alpha = 1$. Player 2 can counterobject with $(0,\beta,\frac{1}{2}-\beta)$ if $\frac{1}{2}-\beta \geq 1-\alpha-\epsilon$, or if $\alpha \geq \beta+\frac{1}{2}$. Hence there is no justified objection of 1 against 2 if

(5) <u>either</u> $\alpha + \gamma = 1$ <u>or</u> $\alpha \geq \beta + \frac{1}{2}$.

By symmetry there is no justified objection of 1 against 3 if

(6) <u>either</u> $\alpha + \beta = 1$ <u>or</u> $\alpha \geq \gamma + \dfrac{1}{2}$.

Now suppose $x = (1,0,0)$. Then (3) is violated; hence $\alpha < 1$. Consider $x = (\alpha, 1 - \alpha, 0)$ with $\alpha < 1$; then (2) is violated. Hence $\alpha + \beta < 1$, and by symmetry $\alpha + \gamma < 1$. So conditions (1) and (2) ensure that $\beta = \gamma$. So any point in the Bargaining Set is of the form $(\alpha, (1 - \alpha)/2, (1 - \alpha)/2)$. Conditions (5) and (6) then imply that $\alpha \geq 2/3$. Hence $\beta + \gamma \leq 1/3$, so that conditions (3) and (4) imply that $\alpha \leq 2/3$. Hence the Bargaining Set consists of the single point $(2/3, 1/6, 1/6)$, as claimed.

References

Aumann, R. J. [1959], "Acceptable Points in General Cooperative n-Person Games," in Contributions to the Theory of Games, vol. IV (Annals of Mathematical Studies, no. 40), A. W. Tucker and R. D. Luce (eds.), Princeton: Princeton University Press, pp. 287-324.

Aumann, R. J. [1961], "The Core of a Cooperative Game without Side Payments," Transactions of the American Mathematical Society, 98, pp. 539-552.

Aumann, R. J. [1967], "A Survey of Cooperative Games without Side Payments," in Essays in Mathematical Economics in Honor of Oskar Morgenstern, M. Shubik (ed.), Princeton: Princeton University Press, pp. 3-27.

Aumann, R. J. and M. Maschler [1966], "Game Theoretic Aspects of Gradual Disarmament," in Development of Utility Theory for Arms Control and Disarmament (Final Report of Contract ACDA/ST-80, prepared by Mathematica Inc., Princeton, N.J. for the U.S. Arms Control and Disarmament Agency), Chapter 5.

Aumann, R. J. and M. Maschler [1967], "Repeated Games with Incomplete Information: A Survey of Recent Results," in Report to the U.S. Arms Control and Disarmament Agency (Final Report on Contract ACDA/ST-116, prepared by Mathematica Inc., Princeton, N.J.) Chapter 3, pp. 287-403.

Aumann, R. J. and M. Maschler [1968], "Repeated Games with Incomplete Information: The Zero Sum Extensive Case," in Report to the U.S. Arms Control and Disarmament Agency (Final Report on Contract ACDA/ST-143, prepared by Mathematica Inc., Princeton, N.J.) Chapter 2, pp. 25-108.

Bewley, T. and E. Kohlberg [1976a], "The Asymptotic Theory of Stochastic Games," Mathematics of Operations Research, 1, pp. 197-208.

Bewley, T. and E. Kohlberg [1976b], " The Asymptotic Solution of a Recursion Equation Occurring in Stochastic Games," Mathematics of Operations Research, 1, pp. 321-336.

Bondareva, O. N. [1962], "Teoriia Idra v Igre n Lits" (The Theory of the Core in an n-Person Game), Vestnik Leningradskogo Universiteta, Seriia Matematika, Mekhanika i Astronomii (Bulletin of Leningrad University, Mathematics, Mechanics and Astronomy Series), no. 13, pp. 141-142 (summary in English).

Bondareva, O. N. [1963], "Nekotorye Primeneniia Metodov Linejnogo Programmirovaniia k Teorii Kooperativnykh Igr" (Some Applications of Linear Programming Methods to the Theory of Games), Problemy Kibernetiki (Problems of Cybernetics), 10, pp. 119-139.

Bott, R. [1953], "Symmetric Solutions to Majority Games," in Contributions to the Theory of Games, vol. II (Annals of Mathematical Studies, no. 28), H. W. Kuhn and A. W. Tucker (eds.), Princeton: Princeton University Press, pp. 319-323.

Davis, M. and M. Maschler [1965], "The Kernel of a Cooperative Game," Naval Research Logistics Quarterly, 12, pp. 223-259.

Gillette, D. [1957], "Stochastic Games with Zero Stop Probabilities," in Contributions to the Theory of Games, vol. III (Annals of Mathematical Studies, no. 39), M. Dresher, A. W. Tucker, and P. Wolfe (eds.), Princeton: Princeton University Press, pp. 179-187.

Kohlberg, E. [1971], "On the Nucleolus of a Characteristic Function Game," SIAM Journal on Applied Mathematics, 20, pp. 62-66.

Kohlberg, E. [1975a], "Optimal Strategies in Repeated Games with Incomplete Information," International Journal of Game Theory, 4, pp. 7-24.

Kohlberg, E. [1975b], "The Information Revealed in Infinitely Repeated Games of Incomplete Information," International Journal of Game Theory, 4, pp. 57-59.

Kurz, M. [1977], "Altruistic Equilibrium," in Economic Progress, Private Values, and Public Policy, North Holland, 1977.

Luce, R. D. and H. Raiffa [1957], Games and Decisions, New York: Wiley.

Maschler, M. and B. Peleg [1966], "A Characterization, Existence Proof and Dimension Bounds for the Kernel of a Game," Pacific Journal of Mathematics, 18, pp. 289-328.

Mertens, J-F. and S. Zamir [1971/72], "The Value of Two-Person Zero-Sum Repeated Games with Lack of Information on Both Sides," International Journal of Game Theory, 1, pp. 39-64.

Nash, J. [1951], "Non-Cooperative Games," Annals of Mathematics, 54, pp. 286-295.

Peleg, B. [1963], "Existence Theorem for the Bargaining Set $M_1^{(i)}$," Bulletin of the American Mathematical Society, 69, pp. 109-110.

Peleg, B. [1967], "Existence Theorem for the Bargaining Set $M_1^{(i)}$," in Essays in Mathematical Economics in Honor of Oskar Morgenstern, M. Shubik (ed.), Princeton: Princeton University Press, pp. 53-56.

Peleg, B. [1968], "On Weights of Constant-Sum Majority Games," SIAM Journal on Applied Mathematics, 16, pp. 527-532.

Schmeidler, D. [1969], "The Nucleolus of a Characteristic Function Game," SIAM Journal on Applied Mathematics, 17, pp. 1163-1170.

Shapley, L. S. [1953a], "A Value for n-Person Games," in Contributions to the Theory of Games, vol. II (Annals of Mathematical Studies, no. 28), H. W. Kuhn and A. W. Tucker (eds.), Princeton: Princeton University Press, pp. 307-317.

Shapley, L. S. [1953b], "Stochastic Games," Proceedings of the National Academy of Sciences, 39, pp. 1095-1100.

Shapley, L. S. [1967], "On Balanced Sets and Cores," Naval Research Logistics Quarterly, 14, pp. 453-460.

Shapley, L. S. and M. Shubik [1969], "On Market Games," Journal of Economic Theory, 1, pp. 9-25.

Stearns, R. E. [1967], "A Formal Information Concept for Games with Incomplete Information," in Report to the U.S. Arms Control and Disarmament Agency (Final Report on Contract ACDA/ST-116, prepared by Mathematica Inc., Princeton, N.J.), Chapter 4, pp. 405-433.

Stearns, R. E. [1968], "Convergent Transfer Schemes for N-Person Games," Transactions of the American Mathematical Society, 134, pp. 449-459.

von Neumann, J. [1928], "Zur Theorie der Gesellschaftsspiele," Mathematische Annalen, 100, pp. 295-320 (translated as "On the Theory of Games of Strategy," in Contributions to the Theory of Games, vol. IV (Annals of Mathematical Studies, no.40, A. W. Tucker and R. D. Luce (eds.), Princeton: Princeton University Press, pp. 13-42).

von Neumann, J. and O. Morgenstern [1944], Theory of Games and Economic Behavior, Princeton: Princeton University Press.

Zamir, S. [1971/72], "On the Relation between Finitely and Infinitely Repeated Games with Incomplete Information," International Journal of Game Theory, 1, pp. 179-198.

Zamir, S. [1973], "On Repeated Games with General Information Function," International Journal of Game Theory, 2, pp. 215-229.

Zermelo, E. [1912], "Über eine Anwendung der Mengenlehre auf die Theorie des Schachspiels," in Proceedings of the Fifth International Congress of Mathematicians, E. W. Hobson and A. E. H. Love (eds.), Cambridge: Cambridge University Press, volume II, pp. 501-504.

Printed and bound by CPI Group (UK) Ltd, Croydon, CR0 4YY

23/10/2024

01778240-0011